CROPS
in POTS

hamlyn

CROPS
in POTS

50 great container projects using vegetables, fruit and herbs

BOB PURNELL
Photography by FREIA TURLAND

An Hachette Livre UK Company
www.hachettelivre.co.uk

First published in Great Britain in 2007 by
Hamlyn, a division of Octopus Publishing Group Ltd
2–4 Heron Quays, London E14 4JP
Copyright © Octopus Publishing Group Ltd 2007
www.octopusbooks.co.uk

ISBN: 978-0-600-61551-4

A CIP catalogue record for this book is available from the
British Library

Printed and bound in China

10 9 8 7 6 5 4

Contents

Getting started   6

How to use this book   8
Why grow edible plants in pots?   9
Siting and grouping containers   11
Mixing edible plants with ornamentals   12
Choosing containers   13
Choosing compost   15
How to plant a pot   16
How to plant a hanging basket   17
Propagation   18
How to grow plants from seed   19
General care   20
Watering   22
Mulching   25
Feeding   26
Pests and diseases   27

Starters   30

Peas in a pod   32
Cut-and-come-again   34
Super salad   36
Tumbling toms   38
Towering thymes   40
Lettuce and tulips   42
Stir-fry   44
Sunny show   46
Lettuce and lobelia   48
Summer cocktail   50
Fishy dish   52
Ornamental onions   54

Main courses   56

Red and gold   58
Potted potager   60
Pods and cobs   62
Perfect partners   64
Floral feast   66
Select salad   68
Sky high   70
Winter vegetable cubes   72
Pasta and pizza pot   74
Once upon a time   76
Textural treats   78
Mint medley   80
On fire!   82
Potato paradise   84
Peas and beans   86
Pretty in purple   88
Roots and shoots   90
Red devil   92
Pepper pot   94
Purple and bronze   96
A taste of the Mediterranean   98
Kale and cabbage   100
On the bay   102
Fireball   104
Green garnish   106
Lots of leaves   108

Desserts   110

An apple a day   112
Blueberry surprise   114
Lemon zest   116
Strawberry ball   118
Gorgeous grapes   120
Currant affair   122
Mellow yellow   124
Tea pot   126
Pop and go   128
Gorgeous gourds   130
Pear delight   132
Passion fashion   134

What to grow   136

Vegetables   138
Fruit   148
Herbs   151
Edible flowers   156

Index   158
Acknowledgements   160

Getting started

How to use this book **8**

Why grow edible plants in pots? **9**

Siting and grouping containers **11**

Mixing edible plants with ornamentals **12**

Choosing containers **13**

Choosing compost **15**

How to plant a pot **16**

How to plant a hanging basket **17**

Propagation **18**

How to grow plants from seed **19**

General care **20**

Watering **22**

Mulching **25**

Feeding **26**

Pests and diseases **27**

How to use this book

This book presents 50 great ideas for growing fruit, vegetables and herbs in containers, divided into Starters (see pages 30–55), Main courses (see pages 56–109) and Desserts (see pages 110–135).

Each project includes a list of all the plants and equipment you will need; step-by-step planting instructions; a beautiful photograph of the finished project to show you how the container will look when it is at its peak; and a delicious recipe that uses one of the plants that you have grown.

The availability of plant varieties changes greatly from year to year, so if you are unable to locate the exact variety given in the following pages, use one with similar attributes.

If you need to know how to care for your container plants, Getting Started (see pages 6–29) provides expert advice on everything you might require, from planting containers and hanging baskets and sowing plants from seed to watering, feeding, mulching and dealing with pests and diseases.

Finally, What to Grow (see pages 136–157) describes the vegetables, fruit, herbs and some edible flowers that are most likely to succeed in containers. So, if you want to try something different, you could adopt an experimental approach and create your own planting scheme by mixing and matching some of the plants in this section.

Key to care symbols

These care symbols are designed to let you know at a glance where to site and how to look after the plants in each project.

 Site in full sun The plants in these projects require as much sun and warmth as possible in order to flourish and crop to their maximum potential.

 Site in full sun or light shade The plants in these projects will grow in either full sun or light shade. Avoid placing them in a particularly hot position as this may cause them to perform less well.

 Requires plenty of watering The plants in these projects are particularly thirsty and require more watering than other projects in order for the fruits to swell and sustain a good crop.

 Requires moderate watering Although they need regular and thorough watering, these plants are more equipped to survive with less water; overwatering may have a detrimental effect on some.

 Feed regularly The plants in these projects are gross feeders and should be given a regular balanced supplementary feed using a liquid or soluble fertilizer to keep them healthy and cropping well.

 Frost hardy The plants in these projects are fully hardy and should survive temperatures as low as -15°C (5°F), although not for prolonged periods. Site containers in as protected a position as possible during cold periods to prevent their roots freezing.

 Frost tender Some of the plants in these projects are frost tender and temperatures below 5°C (41°F) may cause them damage. Temperatutes below freezing (0°C or 32°F) may kill them outright.

 Requires extra care The plants in these projects can be a little temperamental and require a little extra effort and care in order to grow them well and produce a good crop.

Why grow edible plants in pots?

It is hard to imagine a more satisfying feeling than being able to tuck into a plate of food that has been produced by your own hand. Stepping out into your garden and picking fresh ingredients for a meal is a special experience and one that is perfectly possible almost all year round.

With a bit of effort, all manner of vegetables, fruit and herbs can be grown in windowboxes, pots and hanging baskets. And if you're thinking you would rather give over such precious space to plants with more visual appeal, you should think again. As this book demonstrates, there is no reason your edible container garden cannot look just as good as it tastes.

Containers offer you the potential to expand considerably the range of plants you can grow. For example, tender species that might not ordinarily survive the winter outdoors in your garden, such as citrus fruits, can be potted and used to augment the display in summer, then moved to a more protected location for the winter.

Plants that would flounder in your garden soil can be offered a more hospitable home in a container, where you have complete control over the growing conditions. If your soil is alkaline, for instance, yet you yearn to grow blueberries, which require an acidic soil, the answer is to cultivate them in containers filled with ericaceous, or acidic, compost. Equally, containers allow you to grow plants that require lots of moisture, such as courgettes and cucumbers, even though your garden soil may be sandy and quick-draining. If your soil is heavy and slow-draining, containers allow you to grow plants that prefer light, quick-draining soils. In a container you can regulate the watering and customize the compost to suit the plants you are nurturing. In short, when you grow them in containers, you have much more control over your plants and how they perform.

Movable feasts

Containers can also be sited conveniently close to the house, making them easier to manage and care for, and plants in pots are often less likely to be targeted by pests and diseases. If they are attacked, their attackers are far easier to spot and treat because containers are nearer to your eye level.

In this way, plants requiring wildly different soil conditions can be grown harmoniously side by side, albeit in separate pots, and so the opportunities for creating stunning plant partnerships are excitingly broadened. Such plant associations would simply not be possible in the garden border or vegetable plot, but in containers the seemingly impossible can become a reality. By understanding their specific needs and caring for them accordingly, moisture-loving tomatoes and beans can be grown cheek-to-cheek with drought-tolerant herbs, such as thymes; and acid-loving cranberries next to lime-loving cabbage and kale.

Invasive plants such as mint are often best confined to containers where they can be kept firmly under control.

Try growing plants that require different soil conditions side by side. Here a pot of lime-loving cabbages sits next to lime-hating blueberries.

Almost any plant will grow in a container, but wherever possible choose compact or dwarf-growing cultivars.

By the same token, plants that may be too vigorous for the open garden can be kept under control in pots. Mints are a classic example. In containers their roots will be restricted and their vigour sufficiently curbed, but with the correct attention they will thrive. And when they have outgrown their allotted pot, they can be tipped out, divided and repotted with ease.

No-garden gardening
In gardens where there is no soil at all, such as a paved courtyard, roof terrace or balcony, containers can solve all your

A garden isn't essential. Even if you have only a balcony or roof terrace, you can still grow a tasty crop of produce in containers.

problems – unless you prefer to suffer a bleak and depressing outlook, that is. It is in these circumstances that container gardens are at their most vital and, often, their most successful. It is critical not only that they should make the best possible use of space, but they must also sustain all the greenery and flowers required to turn a stark, sterile space into a well-dressed, verdant one.

Containers can also be used in small or awkward areas and, because they are temporary, they can be used to colonize empty areas that may later be required for other purposes.

Above all else, growing edible plants in containers offers you great flexibility and freedom. Generally speaking, containers are easily portable and so you can move, rearrange or regroup them as the need arises. When a plant in a border is in the wrong place you should ideally wait until it is dormant before it can be shifted to a more appropriate position. When it is in a container, the same plant can be moved whenever and, pretty much, wherever you want.

Similarly, pots of plants that have finished cropping can easily be removed and replaced with another container that has been waiting in the wings, and so you can achieve a long succession of tasty crops that are just as pleasing to the eye as to the tastebuds.

Siting and grouping containers

Among the many advantages of growing plants in containers is that you are able to cultivate, side by side, species that require different soil conditions and different levels of feeding. Consequently, it is possible to achieve combinations that would be totally unrealistic in the borders of your garden.

Creating atmosphere

Containers are a great way of brightening up parts of the garden that are otherwise difficult to colonize with plants, such as paved areas or very dry patches at the foot of a high hedge. They can even be used to fill temporary gaps in borders or left in position as permanent fixtures. You can either 'cheat' and hide the pot itself among the other plants or make a full-blown feature of it. Many borders can be improved in this way because the contrast of colour and texture provided by a terracotta pot, for instance, will highlight the plants growing in and around it.

Collections of pots and other containers can be arranged informally or used in a more regimented fashion. A row of potted quarter-standard olives, rosemaries or lavenders in matching pots flanking a flight of steps or symmetrically spaced around a circular or rectangular pool, for instance, will create a strong formal image. Potted herbs, edible flowers and salad vegetables in an assortment of terracotta containers will generate a jumbly, haphazard feel. Indeed, pots can add atmosphere to any setting, and by choosing the appropriate combination of plants, pots and accessories you can achieve any mood you like.

Using space

As a rule, most edible plants thrive in sunny, open positions where they can develop without being drawn towards the light and where fruits will ripen more quickly. Providing shelter from cold winds is advisable for most fruit and vegetables to give them a better chance of yielding a decent crop.

Backgrounds can often be as important as the pots themselves and can contribute considerably to their overall appearance. The right backdrop can make a container look ten times better, while the wrong surroundings may completely nullify its effectiveness. As an example, pots filled with flowers always stand out better if the background is plain and unfussy. In a similar way, variegated plants need a simple background – variegated herbs set against variegated shrubs will look messy and cluttered. As well as other plants, fences, walls and screens all make great backgrounds for plants in containers, but again it is a case of matching the plants to their setting and the setting to the plants.

Finally, you should take advantage of the opportunity that using containers offers in allowing you to site your crops near to seating areas, paths, windows, doors and other places where they can be effortlessly appreciated and readily harvested. Stepping outside the backdoor or leaning from a window to pick a few bits and pieces to use in the evening meal makes it all worthwhile.

By grouping several containers together you can create an attractive but flexible display that can be added to or re-arranged as often as you like.

Mixing edible plants with ornamentals

Although the vast majority of edible plants that are suitable for growing in containers have their own aesthetic appeal, there is absolutely no reason why purely ornamental plants cannot be planted among them in order to achieve an even more attractive display. As a bonus, vegetables are less easily detectable to the many pests that would otherwise swiftly descend upon them, when they are planted among herbaceous perennials and annual flowers,

Striking a balance

The secret of success is to strike a balance between creating a colourful, interesting display and growing a worthwhile crop. Be sparing with ornamental plants and use them to enhance rather than detract from the edible components of the container. Avoid planting vigorous ornamental varieties with slow-maturing, edible plants because they can compete for light, water and nutrients, resulting in a poor or non-existent crop. Always go for ornamental plants that will not interfere with the business of producing a decent crop.

Trailing half-hardy perennials are useful partners as they will fall over the sides of the container as a foil to fruiting and edible-leaved plants that, with a few exceptions, are more upright in their growth. The appearance of fruit trees grown in tubs can be enlivened by sowing hardy annuals, such as dwarf cornflowers (*Centaurea cyanus* cultivars), at their base. As hardy annuals are generally shallow-rooted and can be sown direct

For maximum impact, combine edible plants with colourful ornamentals.

where they are to flower, they will not require planting holes that would damage the tree roots nor are they likely to offer undue competition for food and water as they grow.

Annual climbing plants, such as morning glory (*Ipomoea* species and cultivars), are useful for teaming up with climbing fruits and vegetables. They can, for instance, be allowed to scramble up alongside runner beans or gourds to complement and increase their visual impact.

Beneficial insects

Flowering annuals and perennials planted among crops will attract plenty of the pollinating insects that are essential for fruiting plants, as well as those that are the natural predators of aphids and other common garden pests. Certain plants will even have a hand in deterring less desirable insects. Planting French marigolds (*Tagetes patula*) near tomatoes to repel whitefly is perhaps the most obvious example. This practice, which is known as companion planting, is adopted most often by organic gardeners as a form of pest control.

Take care!

A final word of warning. Even though parts of certain otherwise edible plants are themselves poisonous, take great care not to use poisonous plants as companions for edible ones. They are too numerous to mention here, but an internet search will bring up many lists of plants best avoided or seek advice at your local nursery or garden centre.

Choosing containers

Containers are available in a diverse range of shapes, sizes, styles and materials, from capacious, formal, Italianate marble urns to tiny, plain terracotta flowerpots. Basically, any object that is capable of holding a decent amount of compost and has drainage holes in the base can be used as a container, although if the object is aesthetically pleasing in itself it will make for a more attractive overall display. When it comes to selecting containers in which to grow vegetables and fruit, the general rule is the bigger the better. The larger the pot you provide, the greater and more successful the crop you will reap.

Types of container

It pays to consider a few important factors before you buy. In the first instance you should take into account both the setting and the plants you are intending to grow. Terracotta has universal appeal, suiting any plant and fitting any scene. Metallic containers, on the other hand, are a little more limited in their application and are more likely to lose their appeal in the long term.

Another important issue is practicality. Tall, slender pots are elegant, but unless they are sufficiently weighted they can be top-heavy and are likely to blow over if placed in windy positions. In situations where space is at a premium, such as on a roof garden or balcony, square or rectangular pots, which can be butted together, will make the best use of space. Weight is also a major factor for roof gardens and if you like to move your pots around regularly. In these circumstances lightweight pots made from synthetic materials are a shrewd choice.

Finally, although there are plenty of bargains to be had, it pays to buy the best quality you can afford. Well-made containers usually represent better value in the long term and are an investment that, in many cases, will last a lifetime.

Plastic Although it is not the prettiest material, plastic is among the most practical for pots containing both vegetables and fruit because it is nonporous and so will not dry out as quickly as other materials. It is also lightweight, and plastic containers are relatively easy to move around once planted.

Terracotta The warm, earthy tones of terracotta complement just about any plant. There are dozens of different styles and shapes, from tall, slender cylinders to low, squat troughs, and the colour and finish of terracotta containers will vary depending on where and how they were made.

Because it is porous, terracotta does have a tendency to dry out quickly, but this can be minimized by lining the inside with polythene. Always check that you are buying terracotta that is guaranteed frost resistant.

Glazed Glazed terracotta and stoneware pots are available in a wide range of colours and are often attractively patterned. Most glazed containers available in garden centres have a high degree of frost resistance and will last for years. Considering their long life expectancy, they are relatively inexpensive and are often available with useful matching saucers.

Glazed pots represent good value as they are generally frostproof and so will last indefinitely.

Metal containers are ideal for creating a modern feel but should be lined with polystyrene to insulate plant roots against extremes of temperature.

Stone, reconstituted stone and concrete Natural stone containers are impressive and sculpturally beautiful and provide a distinctive air of permanence, but they are expensive, so less costly substitutes formed from reconstituted stone are a more realistic choice. When weathered, they can look almost the same as the real thing.

Concrete containers have enjoyed a recent revival thanks to their no-nonsense appearance and to the range of textures and finishes that can be achieved.

Containers fashioned of stone, stone substitutes or concrete are highly durable and weather resistant. Their biggest drawback is weight; even the smallest examples are heavy and care should be taken to site them so that they will not have to be repositioned too often.

Terrazzo Available in shades of grey, pink, green and white, terrazzo containers are composed of stone chips set in resin and polished to achieve a marble-like appearance. Although not cheap, they are a cost-effective, long-lasting alternative to

marble, and their clean, smooth lines make them are very much at home in contemporary gardens.

Timber Wood is a sympathetic material for edible plants and for gardens in general. Wooden containers can be affordably custom-built and as such can be relied upon to fit an awkward space. Regular treatment with a plant-friendly wood preservative or varnish will prolong their life and a lick of coloured paint will give them a completely new look.

Metal Modern metal containers are generally best used in contemporary settings and for simple planting schemes, although traditional lead troughs, copper pots and galvanized buckets have been in use for centuries.

In metal containers, more than any other material, plant roots will be subjected to extremes of temperature. On a hot day, in full sun, the metal will heat up considerably, possibly causing damage to delicate roots and drying out the compost. Equally, in winter, metal will provide little insulation against frost. Lining the inner walls with a thin sheet of polystyrene or bubble-wrap polythene should ease the problem.

Fibreglass Fibreglass containers are usually fashioned to mimic other more expensive materials, providing a cost-effective substitute for materials such as lead and stone. The best examples are extremely convincing yet are only a fraction of the price of the genuine article.

Fibreglass containers are lightweight, so are ideal for balconies and similar positions. Although they are weatherproof they are easily damaged if dropped or knocked.

Improvised containers Recycled and reclaimed containers test our inventiveness and imagination. Almost any stable receptacle capable of holding enough compost to sustain plant growth and in which adequate drainage holes can be made is a potential plant container.

Choosing compost

Most garden centres and nurseries offer an extensive array of composts, but essentially there are just three main types, all of which can be adapted as necessary. Never be tempted to use garden soil in place of specially prepared composts because you will risk importing weed seeds, pests and even diseases. Previously used compost may also harbour pests and diseases, so always use fresh.

Soil-based composts
Soil- or loam-based composts are best suited to long-term plants, such as fruit trees and bushes. These composts tend to hold moisture and nutrients for longer than soil-less mixtures do. They are also heavier and more substantial and are therefore useful in windy locations where containers might topple over. However, they do have a tendency to become hard and compacted through watering. In most cases, therefore, they are best lightened by adding a proportion of soil-less compost, perlite or another material that will improve aeration and drainage.

Soil-less composts
Soil-less composts are mainly composed of peat or, increasingly, peat substitutes such as coir. They are readily available, comparatively inexpensive and, consequently, the most widely used. However, they vary enormously in quality, so it is not always wise to opt for the cheapest.

Referred to as general-purpose or multi-purpose composts, they are ideal for temporary plants, such as salad crops, although, with care, longer-term plants will also thrive. Most contain sufficient nutrients for an initial period of six weeks or so, after which it is necessary to add fertilizer, either as granules or in liquid form.

Soil-less composts are lightweight and dry out more rapidly than soil-based composts, so they need to be watered more often. They can also be difficult to re-wet if they are allowed to dry out completely.

Ericaceous composts
Plants that are intolerant of lime and require acidic rather than alkaline soils, such as blueberries and cranberries, must be planted in ericaceous or lime-free compost in order to flourish. Planted in ordinary compost they soon become chlorotic (the leaves turn yellow) and lose the will to live.

Customizing compost
Growing plants in containers affords you the opportunity of catering for their likes and dislikes. As well as enabling you to site them where they will thrive and allowing you to regulate watering and feeding, in containers you can tailor the growing medium to suit the plants by adding various optional ingredients such as grit, where sharp drainage is required, and water-retaining granules to aid moisture retention. Lightweight porous materials such as vermiculite or perlite (both expanded volcanic rocks) will improve soil aeration. The use of these materials will also help to make large containers more easily transportable as well as being useful additives for containers sited on roof gardens and other situations where weight is an issue. They are not usually included as standard components of off-the-shelf composts, but are readily available and easy to incorporate when potting up.

Clockwise from top left: Compost additives (vermiculite, perlite and horticultural grit) and composts (soil-less, ericaceous and soil-based).

How to plant a pot

The basic principles of planting a container are the same no matter what its size, shape or material and no matter what plants are to live in it. Just as thorough preparation pays dividends when planting in the open ground, attention to detail and a little extra effort at the potting-up stage will result in strong, healthy growth. Make sure you have everything you will need to hand before you begin, including clean containers.

It's good practice to wash containers as you empty them to remove any traces of pests and diseases that might be clinging to the sides or base. If the container you are using has been used before, empty out any old compost and clean the inside thoroughly. New or used, make sure that there are adequate drainage holes in the base of your container and, if necessary, drill a few extra.

1 Place pieces of broken terracotta, stones, gravel or, if weight is a factor, chunks of polystyrene in the base to create a drainage layer that will help prevent waterlogging. Partly fill the container with good-quality, fresh compost adapted where necessary to suit the needs of your chosen plants. If it is dry, wet it beforehand because soil-less composts, especially, are difficult to re-wet once completely dried out. Firm the compost down gently as you go and place the largest plant roughly in position to check the level. The top of its rootball should rest about 2.5 cm (1 in) below the rim.

2 While still in their original pots, place all the plants in the container as a trial run, rearranging them until you are happy with the composition. If they are dry, water them thoroughly first. The tallest-growing plants should either go at the back or, if the container is to be viewed from all sides, in the centre. Shorter plants and trailing varieties should be sited towards the front or the edges. Starting with the largest first, tap it out of its pot and plant it, building up the compost level as necessary. Gently tease out the roots of potbound specimens to help them establish more readily.

3 When you are happy with the positioning of each plant, carefully fill around them with more compost and gently firm it down. Avoid firming too heavily because compacted compost will not drain freely or be sufficiently aerated. Make sure that the surface of the compost is roughly level and that there is room between this and the rim to allow for watering and, where appropriate, a layer of mulch. If the compost or mulch finishes level with the top of the container it will spill over the sides when you water.

4 Once planting is complete, water thoroughly to settle the plants in. Remove the rose from the end of the can and direct the water towards the roots. Pour slowly to allow the water to penetrate deeply and so as not to wash out any compost. Replace the rose and sprinkle over the tops of the plants to clean off any splashes of compost. If necessary or desired, apply a decorative mulch as a finishing touch.

How to plant a hanging basket

Even if you do not have or cannot spare any ground space whatsoever, you should be able to find room for one or more hanging baskets or wall planters. From dwarf runner beans to strawberries and a wide range of herbs, you'll be amazed at just what you can grow within the confines of a single basket.

Choose either an open-sided, mesh-type basket that allows you to plant through the sides as well as the top or a solid-sided basket, the sides of which can be disguised by top-planted trailing plants, such as tumbling tomatoes.

Remember that hooks and brackets for suspending hanging baskets must be secure. A well-planted basket filled with moist compost will be very heavy. It is your responsibility to ensure that the hanging basket cannot fall and cause damage or, worse, injury.

1 Rest the basket on an empty pot for stability and place a liner inside. In place of the traditional moss, there are plenty of materials to use including cocoa fibre, synthetic fibre liners and wool, which can be built up in layers. If you use liners made from materials such as polythene, foam or moulded paper pulp you will need to cut holes if you want to plant through the sides.

2 To help retain moisture place an old saucer or a small disc of polythene in the bottom of the basket on top of the lining. Add a little compost and begin pushing plants through the sides from the inside out. At this stage it is important not to damage the roots, which will cause a check in growth.

3 Build up the planting in the basket in two or three layers, adding more plants and compost as you go. If you wish, add water-retaining granules to the compost at this stage. Plant the top of the basket with a more upright-growing plant in the centre and plants that will tumble or trail around the edges. Aim for the level of the compost in the centre of the basket to be slightly lower than that around the edges. This will help to direct water to the roots of the plants.

4 Add controlled-release fertilizer plugs and water thoroughly. Stand the completed basket in a sheltered place to allow the plants to establish for a couple of weeks before hanging up in its final position.

Propagation

One of the great added joys of growing your own food is that most of it can quite easily be raised fairly inexpensively from scratch, from seed or cuttings. Once you are smitten, the pleasure and excitement you gain from raising your own plants seldom fades and is re-awakened with each new batch of seedlings that germinate and every fresh cutting that roots.

Sowing seed

There are few vegetables that cannot be raised easily from seed, especially if you have the small amount of equipment – notably a heated propagating case – necessary to provide high enough temperatures for plants such as tomatoes, peppers and aubergines to germinate. Indeed, so easy are most vegetables to raise from seed that it is hard not to be overrun with them. Most seed packets, except those of F1 hybrid seed, contain far more seeds than you'll need in a single season so don't be tempted to sow the whole packet unless you can handle all the resultant seedlings. As they grow on they will need increasingly more space. It is often worthwhile teaming up with like-minded

---

F1 hybrids

The expression F1 hybrid means the first generation of plants to be derived from the crossing of two distinct, pure-bred lines. F1 hybrids come true to type, and they tend to be uniform and vigorous. F2 hybrids arise when plants from a group of F1 hybrids self-pollinate. They do not necessarily come true.

Substitutions

The plant associations in this book are suggestions only. You can replace them with plants that are available or that you particularly like. Some seed merchants, for example, have their own selections of cut-and-come-again lettuces and oriental greens in a range of leaf shapes, colours and forms, but you can just as easily mix seed from several different packets if you prefer to make your own combination. If you choose other plants always bear in mind that Mediterranean herbs, such as sage and thyme, need a free-draining soil and should not be combined with plants that prefer a heavier, more moisture-retentive medium. Remember, too, that when you are including root vegetables, such as carrots, cultivars that develop a rounded shape can be grown in shallower containers than the more traditional, long carrots.

---

friends and neighbours to put together a combined seed order that can be shared out equally to provide smaller quantities of a wider range of plants.

Some seed, such as that of salad leaves, carrots and beetroot, can be sown directly into the container, although it is often easier to nurture and protect them in the early stages if they are sown in cell or module trays, which are divided into small, individual compartments. These trays also allow for ease of transplanting and avoid root disturbance, which could cause a check in growth. Larger seeds, such as those of melons, cucumbers and squashes are best sown into larger individual pots.

Other methods

Many herbs, such as rosemary, lavender and thyme, will root readily as cuttings, although a few, such as parsley and fennel, are best grown from seed and some others, including all types of mint and lemon balm, can be propagated by dividing them. Every root with a shoot will make a new plant.

Most types of fruit requires more specialist propagating techniques, such as grafting, and because you are unlikely to want huge numbers of any particular type, they are best purchased from a reputable nursery or garden centre.

When you buy make sure that soft fruits, including blackcurrants, gooseberries and strawberries, are certified as virus-free stock.

How to grow plants from seed

Always use fresh, good-quality, specially formulated compost when you are sowing seeds. Formulations for seeds generally contain less fertilizer and are less coarse than those prepared for transplanted seedlings and larger plants. Always follow the specific sowing guidelines on the seed packet, as all seeds need varying conditions and temperatures in order to germinate successfully. For instance, some seeds need light to germinate, while others will not sprout unless all light is excluded.

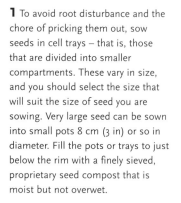

1 To avoid root disturbance and the chore of pricking them out, sow seeds in cell trays – that is, those that are divided into smaller compartments. These vary in size, and you should select the size that will suit the size of seed you are sowing. Very large seed can be sown into small pots 8 cm (3 in) or so in diameter. Fill the pots or trays to just below the rim with a finely sieved, proprietary seed compost that is moist but not overwet.

2 Firm the compost down gently and sow fine seed, such as lettuces, that is often difficult to sow individually in tiny pinches on the surface. They can be thinned out if necessary once germinated. In order to sow larger seed, such as peas, beans and cucumbers, into individual pots, use a pencil, your finger or a dibber to make a hole in the centre that corresponds with the sowing depth suggested on the seed packet and drop in one or two seeds. If both germinate, the weaker of the two seedlings can be discarded.

3 Cover the seed with finely sieved compost, horticultural sand or vermiculite. The sowing depth will vary greatly from variety to variety. Some fine seeds, which need light to germinate, should be simply pressed into the compost and require no covering. Other, relatively large seeds must be buried more deeply. After sowing, water thoroughly using a watering can with a fine rose attached. Place the trays or pots into a cold frame, greenhouse or propagating case, depending on the temperature required for them to germinate. Keep a close eye for signs of germination and make sure they do not dry out. At the same time, be careful not to overwater them, because this can simply cause the seeds to rot.

4 Once the seedlings begin to germinate, allow them maximum light but do not place them in strong, direct sunlight, which may scorch their delicate developing leaves. Frost-hardy species, such as brassicas and lettuce, which are often best sown outside without too much heat, can be planted out into their final containers once they are large enough to handle. Tender peppers, tomatoes and aubergines, on the other hand, must be gently acclimatized and left permanently outdoors only once all danger of frost has passed.

General care

Apart from essential feeding and watering, most container plants will benefit from a little additional routine care. Attention to detail here will pay dividends, making sure that your plants are always looking their best and producing the maximum yield. A major advantage of growing plants in containers is that, because they are nearer to hand and more isolated than plants in the border, you are likely to notice more readily if they are in need of attention or if they are suffering in any way.

Avoiding pests and diseases

General hygiene will cut down the risk of pests and diseases, especially among young and weak plants, which are more prone to infection. Fallen leaves and general plant debris, especially that which has been affected by pests or diseases, should be gathered up and removed on a regular basis. Do not add infected plant material to the compost heap because this will only perpetuate the problem. Bin it instead.

Used containers should be thoroughly cleaned and preferably sterilized before you re-use them, and you must remember to sterilize any pruning equipment that has been used on diseased material. Horticultural disinfectants are widely available.

Pruning and deadheading

Many edible plants, particularly fruiting types, will benefit from a certain amount of pinching out, pruning and training to produce compact, bushy growth and maximum cropping potential. The sideshoots of cordon tomatoes, for instance, should be removed as they appear, but the leading shoots of bush tomatoes may require pinching out to encourage them to bush out fully.

Remove diseased, damaged and dying leaves on a regular basis to keep your plants looking their best and to maintain a good level of hygiene.

Misting over the open flowers of fruiting or podded plants with tepid water will help the fruits set and encourage a more bountiful crop.

Except where plants are being grown for their seed or fruits, remove faded flowers to divert their energies into producing further blooms.

Evergreen herbs grown in pots should be trimmed regularly through the growing season to keep their shape and ensure compact, bushy growth.

Fruit trees in containers require even more careful pruning than those in the ground to encourage the best possible crop. Apples, pears and bush fruits are usually pruned in winter, during the plants' dormancy, although a certain amount of summer pruning may also be necessary. Stone fruits, such as cherries, plums, gages, nectarines and peaches, should be pruned in late spring; never do this in winter. Generally, you should adopt the same pruning regime as if the fruit trees were planted out in the garden, but a little additional judicious pruning may also be required simply to keep them to a manageable size and shape.

Evergreen herbs in pots also need occasional trimming to keep them neat and in proportion, and topiary specimens, such as sweet bay (*Laurus nobilis*) trained as pyramids or lollipops, should be trimmed with hand shears a few times a year to maintain their shape.

Ornamental plants in particular should be deadheaded regularly. Keep an eye on your containers and snip off or pick off all flowerheads as they fade. Once a plant has been allowed to set seed it will tend to stop flowering altogether.

Training and supporting

Many plants will need some form of support system to stop them collapsing under the weight of a heavy crop or rocking around and toppling over in high winds. Climbing plants, such as gourds, runner beans and passionflowers, should be trained on a framework, be it a wooden or metal obelisk or something simpler such as a wigwam of bamboo canes, and tied in regularly to achieve even coverage and good air circulation through the centre of the plants and to avoid a messy tangle of growth. Other plants, such as bush tomatoes and dwarf beans, are also best provided with some support in the form of canes or twiggy prunings pushed in between them.

Wind or tie in the stems of trailing or climbing plants, such as gourds, to prevent them becoming entangled and ungainly.

Watering

Those plants that are established in the garden are largely able to fend for themselves and are free to send their roots through the soil to satisfy their thirst and appetite. Plants in containers, however, cannot access moisture and nutrients beyond the confines of their pot. In effect, a plant in a container can be likened to an animal kept as a pet: it relies on its keeper for all its care. Once you understand this and begin treating your container plants accordingly, you are bound for success. Not that you should become a slave to your containers: gardening, whatever its form, should be enjoyable, not a chore, and while the greater the effort you put in, the richer the rewards, there has to be compromise.

Inserting a plastic tube, with small holes drilled along its length, alongside thirsty plants, such as tomatoes, will make watering your containers easier.

Routine watering

Plants in containers will struggle on indefinitely without food, albeit performing less well and looking increasingly starved. If they are deprived of water for a prolonged period, however, they will inevitably die. A dose of fast-acting fertilizer can buck up the sickliest of plants, but, while even heavily wilted plants can miraculously be revived with a thorough soaking, no amount of water will resurrect a plant that has dried out and withered completely.

Rain cannot always be relied upon to water container plants because the surface of the compost is usually covered by a canopy of foliage that not even the heaviest downpour will penetrate sufficiently. For your plants to flourish, food and water reserves must be replenished manually.

The most efficient way to water is slowly and thoroughly, using a watering can. Remove the rose and direct the water at the roots, not the foliage. Pour on a little and allow it to soak into the compost rather than running off to the sides. Repeat the process from different sides of the pot until the entire rootball is saturated. A length of plastic pipe drilled with small holes and sunk into the pot will make sure that the water goes exactly where it is needed and is especially useful for thirsty plants, such as tomatoes and cucumbers.

Large pots may need 9–14 litres (2–3 gallons) at a time, although they will usually dry out less quickly than smaller pots and so require watering less frequently, although the rate at which they dry out is also governed by the crops they contain.

Small pots can be stood in water-filled saucers to draw up as much moisture as they need, but most plants will resent standing permanently in water so, after a couple of hours, tip away any excess. Containers that have dried out completely can be submerged in a bucket filled with water or in a water butt until the compost is thoroughly soaked.

During summer many containers may need watering twice a day, depending on their position, the weather, the thirst of the individual plant and the type of container. On a hot day a porous terracotta pot in full sun will dry out noticeably faster than a glazed pot in partial shade. Watering in the cool of the morning or evening will reduce the amount of evaporation and avoid scorching delicate leaves or developing fruits.

Most plants will forgive you the occasional missed watering, but if you are apt to forget frequently or are simply pushed for time there are many ways of making watering easier and less time-consuming.

Water-retaining granules

Mixing water-retaining or water-storing granules with the compost at planting time can help considerably. These swell on contact with water, holding moisture and making it available to the roots as required. They are not a substitute for proper watering but will see your plants safely through a missed watering or two.

Water-retaining granules are included in the ingredients for projects when they will be a positive advantage. However, if you find regular watering too much of a chore, you could add a small quantity to any of the containers described in the following chapters.

Simple watering systems are an invaluable time-saver. Drip systems are the most efficient as they direct water to the roots without any wastage.

Wetting agents

Wetting agents work slightly differently and are often included in proprietary, multi-purpose composts, allowing you more easily to re-wet them when they have dried out. Porous pots that dry out quickly, such as those made from terracotta, can be lined inside with polythene to reduce moisture loss, but only line the sides, not the bottom.

If you go away on holiday, try to move as many of your containers as possible to a cool spot. Grouped closely together, they will also help to shade each other.

Watering systems

If you have a large number of containers or you go away frequently, consider investing in a simple watering system. These are inexpensive and are readily available from garden centres. Although, at first glance, they may appear complicated, they are easy to install and highly adaptable.

Drip systems

Drip systems, which release water slowly and efficiently, are the most effective. All you have to do is turn on the tap and walk away before returning hours later to turn it off. Irrigation systems can also be run via a computer attached to a tap, which can be programmed to turn on the water for a set period at a set time every day without your even having to be there.

Overwatering

It is possible to overwater, especially if the container does not drain freely. Waterlogging is a common problem, and the symptoms are similar to those for drought: wilted growth and yellowing or browning leaves that eventually drop. The only way to determine the cause of the problem is to feel the compost and increase or hold off watering as appropriate. Always make sure there are plenty of drainage holes in the base of the container and cover them with a layer of broken crocks to prevent them clogging up with compost. Wherever possible, raise your containers on blocks or special feet to allow free drainage.

Self-watering containers

These miraculous-sounding pieces of gardening equipment are a relatively recent innovation and are becoming increasingly available through garden centres and gardening websites. When used correctly, they work extremely well, and as well as saving you a huge amount of time they invariably help to sustain a better, healthier crop. They are available in a wide range of styles and sizes to suit all types of crop.

Purpose-made self-watering containers tend to be made of plastic so are not always the most aesthetically pleasing, but you can also buy kits that allow you to turn other traditional containers into self-watering ones.

Self-watering containers appear little or no different from any other but their internal anatomy is designed to provide sufficient space for compost and roots with a separate area in the base to hold a reservoir of water. The reservoir is filled via a tube or a hole in the side of the container, which also often acts as an overflow, and the water is drawn up into the compost by capillary action as and when required. In this way the compost remains consistently moist but never waterlogged. As long as the reservoir is kept topped up your plants will never go thirsty.

The mechanics can vary from one type of self-watering container to another. In most instances wicks of capillary matting are used, but others rely on a small area of the compost coming into direct contact with the reservoir. Both methods work well, and some containers employ both. Some form of gauge is also usually included, which allows you accurately to check the water level. You should do so daily if possible, although the water-holding capacity varies depending on the size of the container, and the amount of water used in a day will, of course, depend on weather conditions, the thirst of an individual plant and its siting.

Apart from their labour-saving attributes and ease of use, a huge advantage of self-watering containers is that, because they draw up water from below, nutrients remain available to your plants for far longer than they would if you used traditional watering methods, which tend to wash (leach) nutrients out of the compost. Hence the chore of feeding your container plants is also significantly reduced with self-watering containers.

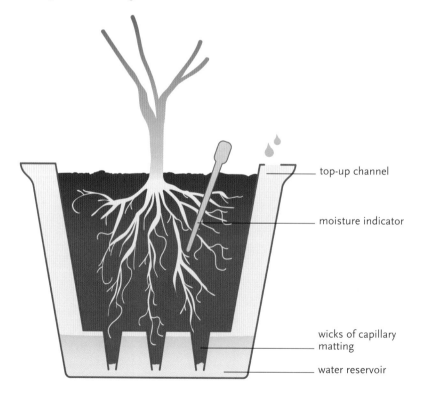

top-up channel

moisture indicator

wicks of capillary matting

water reservoir

Mulching

Any trick that will help conserve moisture in a container is worth adopting, and mulching – covering the surface of the compost with a layer of loose material – is invaluable. In addition to suppressing weeds and helping to keep roots cool and retain moisture in the compost, mulches can also be highly ornamental.

When should I use a mulch?

Mulches are not always necessary. For example, if the surface of the compost will be quickly covered by the plants as they mature, which is often the case with temporary plantings, mulching is generally unnecessary.

In many other cases, however, such as where fruit trees or evergreen herbs, such as sweet bay or lavender, are planted for permanent effect, a mulch applied to a depth of 5–8 cm (2–3 in) will be highly beneficial.

Sympathetically chosen to tie in with the theme of the planting or to complement the style or material of the container, mulches can considerably enhance the overall visual appeal of a container planting.

Materials

A wide variety of materials can be used, ranging from inorganic items such as slate, pebbles and gravel to organic matter that will slowly decompose, including chipped bark, cocoa shells and pine cones.

Slate and pebbles look particularly effective with architectural-looking plants, such as bananas or olives, especially when they are housed in terrazzo, stone containers. Pine cones will enhance the appearance of pots filled with acid-loving blueberries and cranberries.

A layer of sharp grit or crushed seashells around lettuces and cabbages will deter slugs and snails, while around onions and shallots the same material will aid drainage and help to prevent rots developing.

Even recycled materials, such as glass, broken tiles and cracked terracotta pots, can be utilized as mulches, and they can also be combined to great effect – slate with pebbles; glass with marbles; and pine cones with conkers, for example. Indeed, when it comes to being creative with mulches, you can really let your imagination run riot.

Cocoa shells will gradually decompose so should be topped up annually.

A mulch of pebbles can be highly decorative and will last indefinitely.

Feeding

Most proprietary composts contain enough fertilizer to sustain your plants for a maximum of six weeks, after which a weekly or, depending on the crop, even more regular programme of feeding will need to be implemented.

Types of fertilizer

Slow-maturing crops, such as tomatoes, will need a more intense feeding routine to sustain them than will, for instance, fast-maturing cut-and-come again salad leaves, which may require little or no supplementary feeding during their lifespan. Not all vegetables are gross feeders, but all those grown in pots will need more feeding than they would in the open ground.

There are various kinds of proprietary fertilizers, some general, others for more specific application. Controlled-release fertilizers, which last up to six months, can be added in the form of plugs to release nutrients gradually into the compost. Alternatively, slow-release formulations can be added in granule form mixed with the compost when potting up. These are ideal for long-term plantings, such as fruit trees and bushes, but temporary salad crops are best fed with a soluble or liquid fertilizer, which can be diluted and watered into the compost.

Choose a high-potash fertilizer for flowers and fruits. For foliage, select one high in nitrogen. There are also fertilizers formulated for a few specific fruit and vegetables, such as high-potash formulations designed to aid flowering and fruiting of tomatoes – although these are useful for other fruiting plants and a wide range of flowering plants – and compounds for use exclusively on citrus fruits.

A proper regime of feeding will result in healthier plants and, consequently, a better-quality, more bountiful harvest. However, it is actually possible to overfeed and cause damage to your plants, so always check the specific requirements of each crop beforehand and follow the dilution and application rates given on the packaging.

Mix up soluble fertilizers in a watering can or use a special hose attachment, following the manufacturer's recommended dosage.

Controlled or slow-release fertilizer granules can either be mixed with the compost at the potting stage or added afterwards in the form of small plugs.

Pests and diseases

Just as in the garden itself, plants in containers can be affected by a number of pests and diseases. If possible, check every day for signs of infection because the sooner pests and diseases are detected the easier it is to eradicate them. This book describes the most common pests and diseases, but if you need more detailed information you should consult a book or the internet.

Common pests

Aphids Aphids, such as blackfly and greenfly (although there are other colours), are sap-sucking insects that, through feeding, can distort a plant's growth and also transmit viral infections from plant to plant. If discovered early they are easily controlled either by picking off any affected shoots or by spraying with an organic insecticide.

Caterpillars Caterpillars of various kinds, the larvae of a range of insects, can completely devour the leaves of a plant in a matter of hours. The larvae of the cabbage white butterfly can be particularly troublesome on brassicas. In small numbers they can be simply be picked off by hand, but heavy infestations are best dealt with using a contact insecticide.

Leafhoppers These sap-feeding insects live on the undersides of leaves and cause pale, unsightly mottling of the upper leaf surface on a range of plants, including salvias and verbenas. They are tiny green insects, 2–3 mm ($^1/_8$ in) long, with elongated bodies, and they can fly short distances – from leaf to leaf – when disturbed. Spray with a general insecticide or organic aphid control.

Leaf miners The term leaf miner refers to the larvae of insects that mine through the leaf tissue, creating unsightly, straight or more often meandering lines, or occasionally blotches, which usually appear white or grey. Mined leaves usually dry up and die. Remove affected leaves at the first signs of attack and, if necessary, spray with insecticide.

Red spider mite Infestations can be identified mainly by the pale mottling and washed-out appearance of the leaves on which red spider mites are feeding. They are particularly troublesome under glass but can also cause serious and unsightly damage to many outdoor container plants. They are prevalent in hot, dry conditions, and because they multiply continually and rapidly they are difficult to control. Biological control with the tiny predatory mite *Phytoseiulus* is the most reliable combatant, both indoors and out.

Slugs and snails Slugs generally find the majority of their meals at ground level, while snails will climb in search of a tasty morsel. Growing your plants in containers gives you an advantage, but snails will find them eventually. As well as using proprietary slug baits, there are numerous other methods of control. These range from fixing a ring of copper tape just under the rim, which gives them a slight electric shock, to sinking beer or bran traps into the ground around your pots. Slugs supping on beer will quickly become drunk, then fall in and drown. Bran will expand inside them and cause them to explode.

Vine weevils The larvae of vine weevils can cause devastation in containers. The tiny, off-white, brown-headed grubs spend their life eating through the roots of many container plants, which then simply collapse and die. Several effective controls have been introduced in recent years, including both biological nematodes, *Heterorhabditis megidis* or *Steinernema carpocapsae*, and chemical controls, which can be watered into the compost. These controls are expensive and you will need to be persistent to eradicate these pests, but the expense and efforts are worthwhile to ensure that your favourite container plants do not succumb to these voracious pests. Although they do not eat the roots, the adult weevils chew unsightly notches in leaves. Because they are nocturnal, they can be caught at night or provided with an upturned, straw-filled pot to hide out in by day. Check the pot regularly and destroy any weevils you find.

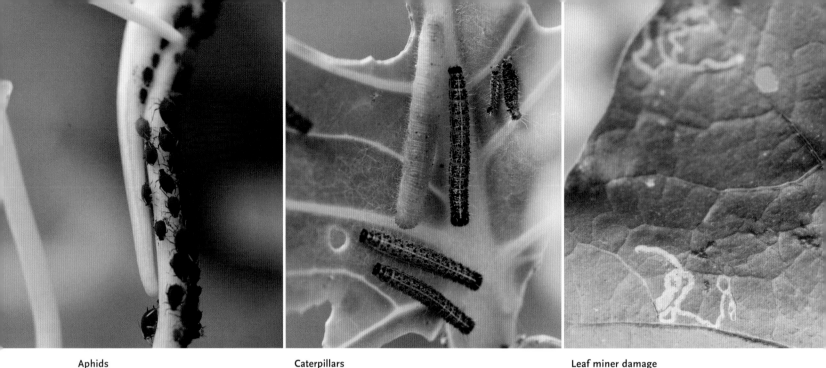

Aphids   Caterpillars   Leaf miner damage

Whitefly Whitefly can affect outdoor plants, especially brassicas and tomatoes, particularly during prolonged periods of hot, dry weather, when they will breed at an astonishing rate. They are sap-feeding insects, about 2 mm (less than ⅛ in) long. Spray infested plants with pyrethrum or an insecticidal soap or use an appropriate insecticide.

Woodlice Large numbers of these distinctive, nocturnal creatures, with their grey, segmented, shell-like bodies, often congregate in, under or around containers. They are generally interested only in decaying plant material and so are not a great threat, but they can damage seedlings and will often attack leaves and stems that have already been partially devoured by slugs and snails. Remove dead leaves and decaying matter regularly from around containers to discourage them.

Common diseases

Blight This fungal disease, which attacks potatoes, tomatoes and related plants, is worse in wet seasons and can devastate a crop. It infects leaves, stems and tubers. Brown patches on the leaves lead to yellowing and premature leaf fall, and in humid conditions the disease will spread rapidly. Spraying with a fungicide from early in the season will help to control the

problem, but the tops of badly affected plants should be removed and destroyed.

Grey mould (botrytis) Botrytis is a fungus that revels in damp conditions and that manifests itself as a furry grey mould, coating leaves and stems. Removing and carefully disposing of affected parts, followed by an application of a general fungicide, may help prevent it from spreading, but seriously affected plants are best thrown away.

Powdery mildew Powdery mildew thrives in dry conditions and shows as white spots, usually on the upper leaf surface, that gradually spread and eventually merge to coat the whole leaf. Affected leaves are unsightly and plants lose vigour. Regular and thorough watering will help prevent the problem, and spraying with a contact fungicide can be worthwhile.

Rust As their name suggests, rusts appear as rusty-brown pustules on the undersides of the leaves of a wide range of plants, including pelargoniums and mints. The upper surfaces of affected leaves are peppered with corresponding yellow spots. Rusts are difficult to control and are best treated as early as possible. Badly affected leaves should be removed

Snails   Whitefly   Powdery mildew

and the rest of the plant sprayed with a fungicide. As a rule, most types of rust are specific to a particular family or genus, so will not necessarily spread to other plants that are growing in the same container.

Controlling pests and diseases

The fact that you are growing your own produce means that you have total control over whether your food is sprayed with garden chemicals. For most of us, the ideal is not to use chemicals at all but instead to employ a range of organic controls. Many gardeners would rather lose a crop than resort to chemicals, but they are available if there is no other option.

Keeping your plants healthy by watering and feeding correctly and practising good hygiene by clearing up plant debris will cut down the risk of attack from pest and disease. In short, prevention is better than cure.

Chemicals There are a number of chemical controls, including insecticides and pesticides, aimed at curing problems on edible plants, but they really should be a last resort. Make sure that you always follow the manufacturer's instructions exactly and never exceed the recommended dosage. Treat chemicals with respect and use them sparingly.

Biological controls Biological controls involve employing the natural enemies of a given pest and are generally effective against only that one pest, so there is no danger of harming other insects that are beneficial. In most cases they are far more effective in the confined environment of a glasshouse or conservatory, but, when correctly applied, nematodes that parasitize vine weevil larvae and soil-dwelling slugs can be used with great success outdoors. Biological controls should never be used in conjunction with chemicals because this will kill the predator as well as the pest.

Other organic controls There are many tried-and-tested methods of combating pests and diseases that do not involve using chemicals. These are loosely termed 'organic controls', and their method and application are many and varied. They range from using plants that deter certain pests – marigolds (*Tagetes*) repel whitefly, for instance – to setting up traps, such as upturned pots that provide cool, dark daytime abodes where slugs and adult vine weevils will congregate so that you can capture and easily dispose of them.

Starters

Peas in a pod **32**

Cut-and-come-again **34**

Super salad **36**

Tumbling toms **38**

Towering thymes **40**

Lettuce and tulips **42**

Stir-fry **44**

Sunny show **46**

Lettuce and lobelia **48**

Summer cocktail **50**

Fishy dish **52**

Ornamental onions **54**

Peas in a pod

Picking peas straight from the pod is twice as pleasant when you can smell the perfume of sweet peas while you do so. This neat pairing combines a diminutive pea variety with a dwarf trailing sweet pea, so both are ideal for even a fairly small container. They will tumble gently over the sides of a hanging basket and are easy to raise from seed so this makes a fun, rewarding project for children. Be sure to pick the sweet peas as often as possible to prolong the display.

Ingredients

1 packet of dwarf pea seed, such as *Pisum sativum* 'Half Pint'

1 packet of dwarf sweet pea seed, such as *Lathyrus odoratus* 'Cupid' (non-edible)

biodegradable fibre pots

seed compost

1 willow hanging basket, 35 cm (14 in) across

multi-purpose soil-less compost

Method

1 Sow both the peas and sweet peas individually into biodegradable fibre pots filled with seed compost and put them in a cold frame to germinate.

2 Select a basket, 35 cm (14 in) across, with closed-in sides (either plastic or a more decorative twiggy design) and fill with soil-less compost.

3 Plant the peas and sweet peas alternately in rings, using 6–8 plants of each to fill the basket. Deadhead the sweet peas and harvest the peas regularly to prolong the display and the crop.

Stir-fried vegetable noodles

Heat 4 tablespoons vegetable oil and stir-fry a bunch of sliced spring onions and 2 thinly sliced carrots. Add 2 crushed garlic cloves, ¼ teaspoon dried chilli flakes, 125 g (4 oz) peas and 125 g (4 oz) halved shiitake mushrooms and continue to stir-fry. Add 3 shredded Chinese leaves and 250 g (8 oz) cooked medium egg noodles. Continue to stir-fry for 2 more minutes; stir through 2 tablespoons light soy sauce and 3 tablespoons hoisin sauce and serve.

Cut-and-come-again

Cut-and-come-again crops are treated exactly as their name suggests: you cut the leaves for salads, usually when they are immature, allow them to regrow and then cut them again. Here we've used a mixture including radish 'Saisai', edible leaf carrot, cress 'Wrinkled Crinkles', kale 'Red Russian' and red amaranth. When sown direct they will germinate quickly and they will be ready to use within weeks.

Ingredients

drainage material (see page 16)

multi-purpose soil-less compost

1 square terracotta container,
  50 x 20 cm (20 x 8 in)

1 bamboo cane

1 packet of mixed salad leaves seed

vermiculite or horticultural sand
  (optional)

Method

1 Put plenty of drainage material in the base of a large, shallow container for drainage and fill to within 5 cm (2 in) of the top with soil-less compost. Water the container thoroughly and allow the compost to drain through.

2 Use a bamboo cane to mark out straight, shallow drills, 8–10 cm (3–4 in) apart, on the surface of the compost.

3 Sow the seed thinly and cover lightly with sieved compost, vermiculite or horticultural sand. Water gently with a fine rose attached to a watering can.

4 If necessary, thin out the germinating seedlings a little to prevent overcrowding and to allow weaker growing varieties to establish.

Pizza bianchi

Put 4 x 20 cm (8 inch) Mediterranean flatbreads on baking sheets and scatter the centres with 200 g (7 oz) crumbled Gorgonzola or dolcelatte cheese. Bake in a preheated oven, 200°C (400°F), Gas Mark 6, for 6–7 minutes or until the cheese has melted and the bases are crisp. Top the pizzas with slices of prosciutto and 50 g (2 oz) salad leaves. Grind over some black pepper and drizzle with olive oil.

Super salad

Contrasting shapes, colours and textures make this group of salad crops as much a treat for the eye as for the tastebuds. The filigree leaves of carrots form a soft centrepiece, while the handsome leaves and stems of beetroot 'Bull's Blood' are a colourful contrast and are, in turn, skirted by bright green, frilly-leaved lettuce 'Fristina', which can be harvested a few leaves at a time.

Ingredients

1 packet of beetroot 'Bull's Blood' seed

1 packet of carrot 'Yellowstone' seed

1 packet of lettuce 'Fristina' seed

seed compost

module trays

1 black, glazed square container, 40 x 35 cm (16 x 14 in)

drainage material (see page 16)

multi-purpose soil-less compost

Method

1 Sow the beetroot, carrots and lettuce in seed compost in module trays to get them started and sow a new batch every few weeks through the spring and summer to keep a succession of vegetables growing.

2 Put plenty of drainage material in the base of the container and fill with a good-quality, soil-less compost.

3 Transfer the seedlings as soon as they are large enough to handle. Space them fairly closely, planting a block of carrots in the centre, surrounded by beetroot with the lettuce round the edges.

Red cabbage and carrot slaw

Make the dressing by whisking together the juice of 1 orange, 1 tablespoon wholegrain mustard, 1 crushed garlic clove and 3 tablespoons olive oil. Season to taste and leave to stand. Meanwhile, finely shred ¼ red cabbage and mix it with 1 sliced red onion, 1 grated carrot, 1 sliced orange pepper and 50 g (2 oz) dry roasted sliced almonds. Spoon over the dressing and serve.

Tumbling toms

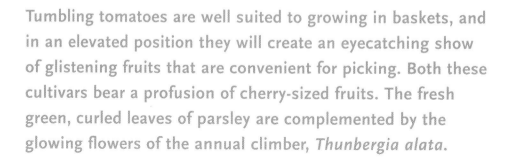

Tumbling tomatoes are well suited to growing in baskets, and in an elevated position they will create an eyecatching show of glistening fruits that are convenient for picking. Both these cultivars bear a profusion of cherry-sized fruits. The fresh green, curled leaves of parsley are complemented by the glowing flowers of the annual climber, *Thunbergia alata*.

Ingredients

1 open-sided wire or wrought iron hanging basket, 35 cm (14 in) across

1 synthetic-fibre hanging basket liner

multi-purpose soil-less compost

water-retaining granules

1 *Petroselinum crispum* (parsley) plant

1 tomato 'Tumbling Tom Red' plant

1 tomato 'Tumbling Tom Yellow' plant

2 *Thunbergia alata* (black-eyed Susan) plants (non-edible)

Method

1 Choose a large open-sided wire or wrought iron hanging basket at least 35 cm (14 in) in diameter and insert a liner. Follow the instructions for planting a hanging basket on page 17.

2 Fill the hanging basket with a good-quality, soil-less compost with added water-retaining granules.

3 Position the parsley in the centre of the basket and a tomato on either side of it. Plant one black-eyed Susan in front of the parsley and one behind it. Leave the basket standing on a pot for a couple of weeks to settle before hanging in its final position.

4 Water at least once a day and feed once a week with a liquid fertilizer to help sustain a worthwhile crop. Pinch out and train the shoots as necessary to maintain an attractive shape.

Tomato and bocconcini bruschetta

Whisk together 3 tablespoons olive oil with 1 teaspoon balsamic vinegar, season to taste and stir in 12 halved bocconcini (baby mozzarella), 20 halved ripe cherry tomatoes and 2 tablespoons chopped basil. Barbecue or toast 4 slices of bread, rub each slice with ½ bruised garlic clove and drizzle over a little more oil. Arrange 125 g (4 oz) rocket over the toast and spoon the tomato and mozzarella mixture on top. Garnish with basil leaves and serve.

Towering thymes

A group of plastic drainage pipes of varying heights and diameters create a funky home for a selection of plants. We've chosen a variety of thymes of differing habits and leaf colours from the grey-leaved, woolly thyme to creamy-white variegated 'Silver Posie'. As a textural contrast, one of the pipes hosts feathery-leaved carrots.

Ingredients

6 plastic drainage pipes of varying height and diameters

silver spray-paint

bricks

6 plant pots of varying depths and diameters (to fit in the pipes)

multi-purpose soil-less compost

horticultural sand

1 packet of carrot 'Red Samurai' seed

1 *Thymus vulgaris* 'Silver Posie' (thyme) plant

1 *Thymus citriodorus* 'Golden King' (lemon-variegated thyme) plant

1 *Thymus pseudolanuginosus* (woolly thyme) plant (non-edible)

1 *Thymus pulegioides* 'Archer's Gold' (thyme) plant

1 *Thymus pulegioides* 'Foxley' (thyme) plant

horticultural grit

Method

1 Spray the drainage pipes with a suitable paint. Stand the bricks on end in the base of each pipe to prop up a plant pot so that the pots will sit inside without showing above the rims of the pipes.

2 Fill a pot with a sandy compost (a mixture of soil-less compost and horticultural sand) to encourage long, straight roots. Sow a few carrot seeds in the top. Thin out the seedlings as necessary once they have germinated.

3 Plant each thyme in a pot in multi-purpose compost mixed with extra grit.

4 Insert the pots into the pipes so that they stand on the bricks.

Parsnip and thyme chips

Thinly slice 2 large parsnips and toss them in a bowl with 1 tablespoon plain flour and 2 teaspoons thyme leaves. Deep-fry the parsnips in batches in vegetable oil for 2 minutes or until they are crisp and golden. Drain on kitchen paper.

Lettuce and tulips

Add an extra dimension to your pots by combining spring-flowering bulbs with colourful early salad leaf crops. The red flowers and mottled foliage of the tulips are set off to perfection by the purple violas and lettuce 'Lollo Rossa'.

Ingredients

9 dwarf tulip bulbs, such as *Tulipa* 'Red Riding Hood' (non-edible)

3 litre (5¼ pint) plastic pot

multi-purpose soil-less compost

1 packet of lettuce 'Lollo Rossa' seed

module trays

seed compost

8 purple viola plants

1 hexagonal terracotta container, 40 x 30 cm (16 x 12 in)

drainage material (see page 16)

Method

1 In autumn plant the tulips in a 3 litre (5¼ pint) plastic pot filled with soil-less compost to start them into growth, ready to transfer to a more decorative pot in the spring.

2 In early spring sow the lettuce in seed compost in module trays and put them in a cold frame to germinate. As soon as the lettuce is ready for planting out, buy eight young viola plants.

3 Put plenty of drainage material in the base of the hexagonal terracotta pot for drainage and fill with a good-quality, soil-less compost. Transfer the tulips to the centre of the pot and plant the violas evenly spaced in a ring around the tulips. Then encircle the violas in turn with seedling lettuce.

4 Deadhead the violas to keep them blooming for as long as possible and pick the lettuce leaves a few at a time, as required.

Red leaf salad with pecan cheese balls

Arrange 175 g (6 oz) lettuce leaves and a handful of edible flowers in a bowl. Add ½ sliced red onion and spoon over a yogurt or sweet mustard dressing. Make the cheese balls by mixing together 250 g (8 oz) goats' cheese or cream cheese with 40 g (1½ oz) finely chopped pecan nuts. Shape the mixture into 16 balls and lightly roll them in paprika. Chill for 20 minutes before serving.

Stir-fry

An inexpensive plastic pot is turned into an appropriate home for this selection of leafy Chinese vegetables by wrapping it with a roll of bamboo border edging. The plants here are grown for their tender young leaves and can be used in salads or stir-fries.

Ingredients

1 packet of pak choi seed, such as 'Mei Qing F1'

1 packet of mizuna greens seed

1 packet of spinach mustard seed

1 packet of red-leaved mustard seed, such as 'Red Giant'

module trays

seed compost

1 square plastic, terracotta or wooden container, 50 x 20 cm (20 x 8 in)

1 roll of bamboo border edging, 2 m (6 ft 6 in) long and 25 cm (10 in) high

plastic-coated wire

drainage material (see page 16)

multi-purpose soil-less compost

Method

1 Sow the seeds in seed compost in module trays, allowing plenty of spares, or sow direct into the container and thin out as necessary. Make successional sowings every few weeks so you have a continuous supply of leaves.

2 Use a plain, shallow, square pot and wrap around it a length of bamboo border edging, which is available from garden centres. Fasten the ends together with plastic-coated wire. Put plenty of drainage material in the base and fill with soil-less compost.

3 When they are large enough to handle, plant out the seedlings in rows to create a striped effect. Repeat each variety a couple of times to create a more even and interesting spread of leaf colour. Water well during dry spells.

Sesame prawns with pak choi

Mix together 1 teaspoon sesame oil, 2 tablespoons light soy sauce, 1 tablespoon honey, 1 teaspoon grated fresh root ginger, 1 teaspoon crushed garlic and 1 tablespoon lemon juice and use the mixture to marinate 600 g (1 lb 3 oz) raw peeled tiger prawns (tails left on) for 5–10 minutes. Cut 500 g (1 lb) pak choi in half, blanch for 40–50 seconds and drain. Heat 2 tablespoons vegetable oil and cook the prawns and marinade for 3–4 minutes. Arrange the pak choi on plates with the prawns and pan juices on top.

Sunny show

Brightly coloured dwarf sunflowers jostle with low-growing, red- and white-flowered runner beans in this lively planting. There are many varieties of dwarf sunflowers growing 45–60 cm (18–24 in) high, including 'Dwarf Yellow Spray', which has many relatively small heads on each plant. In front is runner bean 'Hestia', which will grow to the same height as the sunflowers but has a more relaxed habit.

Ingredients

1 packet of dwarf runner bean 'Hestia' seed

3 dwarf sunflower plants, such as 'Dwarf Yellow Spray', or a packet of seed

biodegradable fibre pots

seed compost

1 lead-effect fibreglass trough, 60 x 20 cm (24 x 8 in)

drainage material (see page 16)

multi-purpose soil-less compost

well-rotted garden compost or manure

6 twiggy sticks, 30–45 cm (12–18 in) long, to support runner beans

soluble fertilizer

Method

1 Sow the beans and sunflowers into seed compost in individual biodegradable fibre pots that will slowly disintegrate and avoid root disturbance.

2 Make sure there are drainage holes in the trough. Put plenty of drainage material in the base of the trough and fill it with a good-quality, soil-less compost with added organic matter, such as well-rotted garden compost or manure, for moisture retention.

3 When they are large enough, plant out the sunflowers, spacing them evenly along the back of the trough, with five or six runner beans along the front edge. Push twiggy sticks into the compost to provide support for the beans and sunflowers. Water well and feed occasionally with a soluble fertilizer.

Stir-fried hoisin beans

Blanch 500 g (1 lb) beans in boiling water for 2 minutes, then drain well. Heat 2 tablespoons of vegetable oil in a wok, add 2 sliced cloves of garlic and 2 red chillies that have been deseeded and sliced. Stir briefly, then add the beans, 6 tablespoons of hoisin sauce and 1 teaspoon of salt to the wok. Stir-fry over a high heat for 1–2 minutes until the beans are tender. Serve immediately with plain boiled rice or noodles.

Lettuce and lobelia

Few leafy vegetables are as versatile, easy-to-grow and as visually appealing as lettuce. They can be grown in baskets, windowboxes and all manner of other containers and either allowed to mature and be harvested whole or, in the case of many varieties, treated as a cut-and-come-again crop. Interplanting the lettuce with non-edible, pale blue-flowered trailing annual lobelia makes for a striking contrast.

Ingredients

1 packet of mixed coloured-leafed lettuce seed

seed compost

module trays

15 pale blue trailing lobelia plants (non-edible)

1 wrought iron or plastic-coated metal half-basket, 40 cm (16 in) across

hanging basket liner

multi-purpose soil-less compost

Method

1 Sow the lettuce in tiny pinches in seed compost in module trays in mid- to late spring and place them in a cold frame or sheltered place where they will germinate readily. Overcrowded bunches of seedlings should be thinned down to one strong plant.

2 Buy the lobelia plants from a nursery or garden centre once the lettuces are large enough to plant out. Choose a large, hayrack-style half-basket, 40 cm (16 in) across, and insert a liner. Using soil-less compost, follow the instructions for planting a hanging basket on page 17.

3 Make sure that the plants are evenly distributed over the sides and top of the basket and water well. Fix the basket to a wall or fence in a sunny but not scorchingly hot position. Keep well watered to prevent the lettuce from bolting.

Thai-dressed tofu rolls

Put 8 lettuce leaves in boiling water for 10 seconds. Rinse and drain them. Shred 16 more lettuce leaves and mix with 275 g (9 oz) diced tofu and 100 g (3½ oz) shredded mangetout. Mix together 2 tablespoons sesame oil, 2 tablespoons soy sauce, 2 tablespoons lime juice, 1 tablespoon muscovado sugar, 1 sliced Thai chilli and 1 crushed garlic clove. Mix the sauce and tofu mixture and spoon into the lettuce leaves. Roll up and chill.

Summer cocktail

What better on a warm summer evening than to be able to pick fresh flowers, fruits or leaves to add a refreshing taste to a summer beverage? Tiny alpine strawberries, borage and salad burnet are among the best for this purpose, and a galvanized steel bucket is an unusual but somehow fitting home to them.

Ingredients

1 galvanized metal bucket, 30 cm (12 in) across

drainage material (see page 16)

multi-purpose soil-less compost

slow-release fertilizer

water-retaining granules

1 *Borago officinalis* (borage) plant

3 *Sanguisorba minor* (salad burnet) plants

4 alpine strawberry plants

Method

1 Drill several large holes in the bottom of the bucket for drainage, wearing goggles to protect your eyes from metal splinters. Then drop in a layer of drainage material.

2 Fill the bucket with good-quality, soil-less compost mixed with a few slow-release fertilizer granules and water-retaining granules.

3 Plant the borage in the centre of the container, the salad burnet around it, and the strawberries around the edge so the fruit will drape over the rim.

Strawberry lemonade

Put 10 strawberries, 10 mint leaves and 100 ml (4 fl oz) sugar syrup in a blender or food processor and blend to a purée. Transfer the purée to a large jug, add 100 ml (4 fl oz) fresh lemon juice, 1 litre (1¾ pints) soda water and plenty of ice cubes and stir well. Serve in goblets, decorated with strawberry slices, sprigs of mint and borage flowers.

Fishy dish

Handy for the kitchen, this selection of herbs will provide a summer-long supply of fresh leaves that can be used in a variety of recipes, especially fish dishes. French tarragon is a versatile herb with a strong, distinctive flavour that complements chicken, veal and rice dishes, as well as fish.

Ingredients

1 wooden windowbox, 60 x 20 cm (24 x 8 in)

drainage material (see page 16)

multi-purpose soil-less compost

horticultural grit

1 *Artemisia dracunculus* (French tarragon) plant

2 *Anethum graveolens* (dill) plants; choose a dwarfer variety, such as 'Bouquet'

2 *Petroselinum crispum* (parsley) plants

3 *Thymus citriodorus* 'Golden King' (lemon-variegated thyme) plants

2 *Thymus vulgaris* 'Silver Posie' (thyme) plants

Method

1 Make sure there are adequate drainage holes in the base of the windowbox. Add a layer of drainage material and fill with good-quality, soil-less compost with added grit so that it is drains freely.

2 Plant the tarragon at the back in the centre with the dill on either side and the parsley at each end. Alternate the thymes in a row in front where they will gently tumble over the edge of the trough.

3 Harvest the leaves a few at a time, as and when you need them. After flowering, trim back the thyme to keep it bushy and encourage a fresh new crop of leaves. Take care not to overwater this container because the tarragon, especially, will resent being overwet.

Chicken and tarragon salad

Put a 1.5 kg (3 lb) chicken in a saucepan with 1 sliced onion, the juice and rind of 1 orange, 1 tablespoon chopped tarragon and 1 bay leaf. Cover with water and simmer for 45–60 minutes. When the chicken is cooked and cool, cut it into pieces. Mix 300 ml (½ pint) of the cooled-down stock with 1 tablespoon olive oil and 1 tablespoon white wine vinegar and pour over the chicken. Garnish with oranges, cress and sprigs of tarragon.

Ornamental onions

Salad or spring onions are partnered by spinach in this old wooden crate to provide several weeks' worth of tasty salad ingredients, utilizing every square inch of space. Spinach is a highly nutritious crop packed with iron and vitamins and can either be lightly steamed or the fresh young leaves, particularly, used raw in salads.

Ingredients

1 packet of onion 'Red Beard' (or similar) seed

1 packet of onion 'Shimonita' (or similar) seed

module trays

seed compost

1 wooden crate, 45 x 30 cm (18 x 12 in)

polythene sheet

drainage material (see page 16)

multi-purpose soil-less compost

1 packet of spinach 'Bordeaux' seed

Method

1 Sow the onions in seed compost in module trays, one or two seeds per compartment and place in a cold frame to germinate.

2 Line the wooden crate with polythene and pierce plenty of holes in the base for drainage. Add a layer of drainage material and fill with good-quality, soil-less compost. Once they are large enough to handle, plant the onions in rows. Initially they can be quite closely spaced, then gradually thinned out, with the thinnings being put to good use in salads.

3 Leave the remaining onions to mature for use later, and sow spinach direct into the compost between them. Cover lightly with compost. The spinach will germinate rapidly and should be harvested as baby leaves to avoid affecting the development of the onions.

Pear, Stilton and spinach salad

Core and thickly slice 4 pears and cook the slices for 1 minute on each side on a hot griddle. Sprinkle them with 4 tablespoons lemon juice. Pile 250 g (8 oz) baby spinach on a large plate and arrange the pear slices on top. Sprinkle 4 chopped walnuts and 250 g (8 oz) crumbled Stilton cheese on top. Spoon 4 tablespoons walnut oil over the salad and serve.

Main courses

Red and gold **58**

Potted potager **60**

Pods and cobs **62**

Perfect partners **64**

Floral feast **66**

Select salad **68**

Sky high **70**

Winter vegetable cubes **72**

Pasta and pizza pot **74**

Once upon a time **76**

Textural treats **78**

Mint medley **80**

On fire! **82**

Potato paradise **84**

Peas and beans **86**

Pretty in purple **88**

Roots and shoots **90**

Red devil **92**

Pepper pot **94**

Purple and bronze **96**

A taste of the Mediterranean **98**

Kale and cabbage **100**

On the bay **102**

Fireball **104**

Green garnish **106**

Lots of leaves **108**

Red and gold

The small, round, golden-yellow courgette 'One Ball F1' is a compact, prolific variety and a good contrast with the orange-red-fruited bush tomato 'Czech's Bush'. These are both thirsty, greedy plants, so water and feed regularly for a good, sustained crop.

Ingredients

2 courgette 'One Ball F1' plants or a packet of seeds

small plastic plant pots

seed compost

1 glazed cylindrical container, 45 x 40 cm (18 x 16 in), neutral brown or green

1 self-watering container kit, 40 cm (16 in) across (see page 24)

multi-purpose soil-less compost

1 tomato 'Czech's Bush' plant or similar bush variety

3–4 bamboo canes, 1.2 m (4 ft) long

Method

1 If you are growing from seed, sow the courgettes in seed compost in small individual pots in mid- to late spring and place in a heated propagator. Alternatively, sow them in an unheated greenhouse once all danger of frost has passed.

2 Set up the self-watering container kit inside the glazed container, fill with good-quality, soil-less compost and plant the tomato and courgettes in a triangle. Water well and feed regularly. The tomato may need a bamboo cane or two to support its weight when it is laden with fruit. Pinch out the tip of the tomato plant to encourage it to bush out.

3 Regularly remove a proportion of the oldest courgette leaves to allow air and sunlight through to ripen the fruit and prevent the tomato from being swamped. Pick the courgettes when they are just 7.5 cm (3 in) across to produce a season-long succession of fruits.

Linguine with summer vegetables

Skin and slice 1 red pepper. Slice 1 courgette, 1 red onion and 1 small aubergine. Trim 8 asparagus spears and cook all the vegetables on a hot griddle. Transfer to a dish and add 3 tablespoons cooked peas. Drizzle with 5 tablespoons olive oil and keep warm while you cook 300 g (10 oz) linguine. Mix the cooked pasta and vegetables together, season to taste and scatter with grated Parmesan cheese. Garnish with torn basil leaves and serve.

Potted potager

A mouth-watering mixture of vegetables, edible flowers and herbs will provide produce all summer long. Housing them in a deep, generous-sized container makes it easy to grow a wide range of produce in a small space.

Ingredients

1 packet of compact-growing kale (such as 'Dwarf Green Curled') seed

1 packet of *Tropaeolum majus* (nasturtium) seed

1 packet of beet 'Bull's Blood' seed

seed compost

module trays

1 square wooden container, 75 x 45 cm (30 x 18 in)

sheets of polythene

drainage material (see page 16)

multi-purpose soil-less compost

soil-based potting compost

1 *Thymus vulgaris* 'Silver Posie' (thyme) plant

1 *Petroselinum crispum* (parsley) plant

2 perpetual-fruiting strawberry plants

1 *Coriandrum sativum* (coriander) plant

1 *Anethum graveolens* (dill) plant

1 *Allium tuberosum* (garlic or Chinese chives) plants

6 purple and white viola plants

Method

1 In spring, sow the various seeds in seed compost in module trays and put them in a cold frame to germinate. You will need only a few plants of each variety, so don't oversow.

2 Position the wooden container before lining it and filling it with compost and plants, because it will be too heavy to move afterwards. Line the sides of the container with polythene to prolong its life (but do not cover the drainage holes). Place a layer of drainage material in the base and fill with a half-and-half mix of soil-less and soil-based compost.

3 Once the seedlings are ready for planting out, purchase the other plants. Position the bought plants first and plant the seedlings in small informal blocks between them. Taller varieties, such as coriander, should be planted towards the centre of the container and trailing or low-growing plants, such as strawberries and thyme, should be planted towards the edge.

Falafel cakes

Blend 400 g (13 oz) canned chickpeas with 1 onion, 3 garlic cloves, 2 teaspoons cumin seeds, 1 teaspoon mild chilli powder, 2 tablespoons chopped garlic chives, 2 tablespoons chopped mint, 3 tablespoons chopped coriander and 50 g (2 oz) breadcrumbs. Season to taste. Flatten spoonfuls of the mixture into cakes and fry in batches in 1 cm (½ inch) vegetable oil for about 3 minutes, turning once, until golden.

Pods and cobs

A half-barrel is needed for tall, thirsty, fast-growing crops such as sweetcorn. 'Minipop F1' has been bred to be harvested as a mini vegetable, producing succulent 'baby' cobs. An underplanting of dwarf French beans covers and shades the soil at the base of the corn.

Ingredients

1 packet of sweetcorn 'Minipop F1' seed

1 packet of dwarf French bean 'Purple Teepee' seed

1 packet of dwarf French bean 'Golden Teepee' seed

small plastic plant pots

seed compost

1 wooden half-barrel,
  70 x 40 cm (28 x 16 in)

drainage material (see page 16)

multi-purpose soil-less compost

well-rotted garden compost or manure

Method

1 Sow the sweetcorn and beans in seed compost in individual pots in mid-spring and put them in a heated propagator or greenhouse to germinate. Alternatively, sow them later in spring in a frost-free greenhouse or cold frame.

2 Drill plenty of drainage holes in the base of the barrel, cover them with a layer of drainage material and fill with a good-quality, soil-less compost. Mix in additional organic matter, such as well-rotted garden compost or manure, to improve moisture retention. Position the container where it will receive as much sun as possible.

3 Plant five or six evenly spaced sweetcorn around the container and plant six of each type of bean between them. Harvest the beans regularly to keep them cropping and pick the corn as mini cobs before the grains begin to swell. Keep well watered and feed regularly.

Grilled sweetcorn with chilli and lime

Remove the husks from 4 corn cobs. Mix together 1 tablespoon coarse chilli powder and 1 tablespoon sea salt. Cook the cobs under a grill for 4–5 minutes, turning them regularly so they cook all over and are slightly charred in places. Dip half a lime in the chilli mixture and squeeze it over the cobs. Repeat with 3 more halves of lime to coat the cobs and serve immediately.

Perfect partners

These containers combine edible plants with those that are purely ornamental to produce flowers and crops all summer long. Each of the three tall, circular containers is sown with a single crop that can be harvested as mini-veg or young leaves and replaced halfway through the season or left to mature. Each crop is surrounded by tender perennial trailing plants to create a set of stunning pairings.

Ingredients

1 packet of lettuce 'Revolution' seed

1 packet of leek 'Armor' seed

1 packet of beetroot 'Kestrel F1' seed

module trays

seed compost

1 round zinc, silver-coloured container, 35 x 30 cm (14 x 12 in)

1 round zinc, silver-coloured container, 40 x 35 cm (16 x 14 in)

1 round zinc, silver-coloured container, 50 x 40 cm (20 x 16 in)

drainage material (see page 16)

multi-purpose soil-less compost

4 *Plecostachys serpyllifolia* (non-edible)

5 *Verbena* Aztec Magic Plum (non-edible)

4 *Sutera cordata* Lavender Showers ('Sunlav') (non-edible)

Method

1 Sow the vegetables in seed compost in module trays in mid- to late spring. You will need only a few for each container, so don't sow too many unless you need them elsewhere.

2 If necessary, drill holes in the base of each container for drainage and add a layer of drainage material before filling with soil-less compost.

3 Plant three lettuce surrounded by plecostachys in the largest container, and three or four leeks skirted by verbena in the medium-sized pot. In the smallest container plant a central block of five beetroot, with the sutera planted around the edge. Have a few later-sown plants on hand to replace the crops as they are harvested, because the tender perennials will continue until the frosts.

Rocket, beetroot and red pepper salad

Cut 4 cooked beetroot into 6–8 pieces, drizzle with olive oil and roast in a preheated oven, 200°C (400°F), Gas Mark 6, for 30 minutes. Blend 25 g (1 oz) roasted hazelnuts with 1 tablespoon balsamic vinegar, 1 garlic clove and 4 tablespoons olive or hazelnut oil. Arrange 75 g (3 oz) rocket leaves, 300 g (10 oz) chargrilled red peppers and the beetroot on a plate. Drizzle over the dressing.

Floral feast

A surprising number of flowers are edible. This summery feast of edible flowers includes pot marigolds, violas, nasturtiums, cornflowers and sunflowers.

Ingredients

1 packet of *Helianthus annuus* 'Teddy Bear' (sunflower) seed

1 packet of *Calendula officinalis* 'Pink Surprise' (pot marigold) seed

1 packet of *Centaurea cyanus* 'Florence blue' (dwarf cornflower) seed

1 packet of *Tropaeolum majus* 'Moonlight' (nasturtium) seed

1 packet of *Tropaeolum majus* 'Empress of India' (nasturtium) seed

module trays

seed compost

selection of terracotta plant pots of
varying shapes and sizes, including
1 tall pot 35 x 45 cm (14 x 18 in),
1 trough 50 x 20 cm (20 x 8 in),
2 pots 35 x 25 cm (14 x 10 in) and
1 small pot 10 x 10 cm (4 x 4 in)

drainage material (see page 16)

multi-purpose soil-less compost

6 purple and white viola plants

6 yellow viola plants

1 *Lavandula angustifolila* 'Princess Blue' (lavender) plant

6 pale blue and cream viola plants

Method

1 Sow the various seeds in seed compost in module trays in mid-spring and put them in a cold frame. When the seedlings are large enough for transplanting, buy the lavender and viola plants ready for planting.

2 Put plenty of drainage material in the bases of the terracotta pots and fill them with soil-less compost. Plant three sunflowers in a tall pot surrounded by purple and white violas, six pot marigolds surrounded by a ring of seven or eight cornflowers in a bowl, the nasturtiums and yellow violas in a trough, the lavender as a single specimen plant and the three pale blue and cream violas in a small pot.

3 If possible, have a couple of spare pots of later-sown plants waiting in the wings to replace any of the original plantings once they are past their best. Deadhead regularly so the plants do not divert their energies into seed rather than flower production.

Cottage garden salad

Put about 250 g (8 oz) torn mixed salad leaves, such as rocket, escarole, red oakleaf, salad burnet, frisé, radicchio and lamb's lettuce, into a salad bowl with a handful of fresh herb sprigs, with flowers, such as nasturtiums, fennel, chives, dill and mint. Hull and halve 250 g (8 oz) small strawberries and add them to the salad bowl. Season to taste, spoon over a yogurt or orange dressing and serve.

Select salad

Striking edible leaves and stunning blooms combine to create this colourful collection of unusual salad ingredients. The rich green, highly nutritional leaves of parsley are invaluable for garnishing, as well as enhancing the flavour of other foods.

Ingredients

1 packet of *Calendula officinalis* 'Greenheart Orange' (pot marigold) seed

1 packet of *Rumex acetosa* blood-veined (blood-veined sorrel) seed

1 packet of *Barbarea vulgaris* 'Variegata' (variegated American or land cress) seed

module trays

seed compost

1 lead-effect fibreglass container, 30 x 30 cm (12 x 12 in)

drainage material (see page 16)

multi-purpose soil-less compost

2 *Petroselinum crispum* (parsley) plants

Method

1 Sow the various seeds in seed compost in module trays in a cold frame in spring and move to a sheltered spot outdoors once the plants have germinated and are growing strongly.

2 When the seedlings are large enough, plant them out into a container, adding a layer of drainage material to the container before filling with soil-less compost. Buy two young parsley plants to accompany them. Plant one pot marigold in the centre and surround with a ring of alternating parsley, sorrel and land cress, using two plants of each.

3 Pick leaves regularly, whether you intend to use them or not, because this will encourage fresh, tasty new growth. Remove faded pot marigold blooms to prevent them from running to seed.

Herb omelette

Mix 1 tablespoon wholegrain mustard with 40 g (1½ oz) unsalted butter and spread the mixture over the undersides of 4 flat mushrooms. Cook under a grill for 5–6 minutes. Meanwhile, beat 2 tablespoons chopped mixed herbs (such as chives, parsley and tarragon) with 4 eggs and season to taste. Melt about 15 g (½ oz) butter in a nonstick frying pan, swirl in the egg mixture and cook. Slide the omelette on to a warmed plate, add the mushrooms and serve.

Sky high

Planted in a capacious wooden half-barrel and clambering up a wigwam, golden-leaved, scarlet-flowered runner bean 'Sun Bright' is as ornamental a climber as you could wish for. For an extra splash of colour we've added blue-flowered morning glory to create a truly breathtaking combination.

Ingredients

1 packet of runner bean 'Sun Bright' seed

seed compost

small plastic plant pots

1 packet of *Ipomoea* 'Flying Saucers' (morning glory) seed (or a similar cultivar) (non-edible)

module trays

1 wooden half-barrel, 70 x 40 cm (28 x 16 in)

drainage material (see page 16)

multi-purpose soil-less compost

1 willow obelisk, 1.5 m (5 ft) high, or 6 bamboo canes, each 1.8 m (6 ft) long

Method

1 Sow the runner beans in seed compost in individual pots and the morning glory in module trays in spring. Place in a heated propagating case or heated greenhouse to germinate. When they are large enough, pot the morning glory on into large pots.

2 Drill plenty of drainage holes in the base of the barrel and cover with a layer of drainage material. Fill with good-quality, soil-less compost. Plant out the beans and the morning glory once all danger of frost has passed. A half-barrel should accommodate five of each, planted alternately.

3 Provide support in the form of a willow obelisk or a wigwam of bamboo canes, at least 1.5 m (5 ft) high, and wind in the shoots to stop them becoming too tangled. Feed and water regularly.

Green vegetable risotto

Heat 50 g (2 oz) butter and 1 tablespoon olive oil and sauté a crushed garlic clove and a chopped onion. Add 300 g (10 oz) arborio rice and stir to combine, then gradually add 1 litre (1¾ pints) hot stock and stir continually, until it has been absorbed. Add 125 g (4 oz) each of trimmed runner beans, peas, broad beans, asparagus and spinach, and stir in 75 ml (3 fl oz) dry vermouth or white wine. Cook, stirring, for another 2 minutes, then stir in 50 g (2 oz) butter, 2 tablespoons chopped parsley and grated Parmesan.

Winter vegetable cubes

These handsome containers play host to a trio of good-looking vegetables, planted in blocks to create a stunning autumn and winter display. Once harvested, the cauliflower trough can be moved away, leaving the chard and kale to continue looking stunning throughout winter. The mulch not only looks great but will also help conserve moisture.

Ingredients

1 packet of cauliflower 'Candid Charm F1' seed

1 packet of kale 'Redbor F1' seed

1 packet of red-stemmed chard or leaf beet seed

module trays

seed compost

1 square, charcoal-grey, terrazzo, reconstituted-stone container, 40 x 40 cm (16 x 16 in)

1 square, charcoal-grey, terrazzo, reconstituted-stone container, 35 x 35 cm (14 x 14 in)

1 charcoal-grey, terrazzo, reconstituted-stone trough, 50 x 30 cm (20 x 12 in)

drainage material (see page 16)

multi-purpose soil-less compost

slate chippings or polished pebbles

Method

1 Sow the seeds in late spring or early summer in seed compost in module trays for an autumn or early-winter crop. The kale can be harvested as baby leaves, so make successional sowing through the first half of summer, allowing the final batch of seedlings to mature fully.

2 When the seedlings are large enough, plant the kale in the largest container in a block of nine, the chard or leaf beet in the smaller square container, again in a block of nine, and the cauliflower in a row of three in the trough. Using soil-less compost, follow the instructions for planting a container on page 16.

3 Mulch around each group of plants with slate or pebbles and keep well watered. Be vigilant for caterpillars and pick them off immediately.

Cauliflower tarka

Cook 1 teaspoon cumin seeds and 1 teaspoon yellow mustard seeds in 1 tablespoon sunflower oil. Add 500 g (1 lb) cauliflower florets, 2 finely chopped garlic cloves, 2.5 cm (1 inch) fresh root ginger, shredded, 2 sliced red chillies and stir-fry for 6–7 minutes. Stir in ½ teaspoon garam masala and 200 ml (7 fl oz) hot water. Cover and cook on high for 1–2 minutes. Season to taste and serve.

Pasta and pizza pot

A dwarf bush tomato plant, which will yield a hefty crop of tasty fruits, can be surrounded by a mixture of marjorams with various leaf colours, including gold-tipped, golden curly and variegated forms, and oregano to add colour and texture. Annual rocket, which has finely cut leaves and attractive spikes of pale yellow flowers, completes the planting.

Ingredients

1 lead-effect fibreglass container, 40 x 40 cm (16 x 16 in)

drainage material (see page 16)

multi-purpose soil-less compost

1 tomato 'Totem' plant

1–2 bamboo canes, about 1.2 m (4 ft) long

1 *Origanum vulgare* (oregano or wild marjoram) plant

1 *Origanum vulgare* 'Aureum Crispum' (golden curly marjoram) plant

1 *Origanum vulgare* 'Aureum' (golden marjoram) plant

1 *Origanum vulgare* 'Country Cream' (variegated marjoram) plant

1 *Origanum vulgare* 'Gold Tip' (gold-tipped marjoram)

1 packet of cut-leaved *Eruca vesicaria* (annual rocket) seed

Method

1 Make sure there are plenty of drainage holes in the base of the container. Cover them with a layer of drainage material and fill with soil-less compost.

2 Plant the tomato in the centre of the pot and push in a couple of bamboo canes for support. Evenly space the various marjorams around the tomato near the edge of the pot so they will tumble forward and obscure the straight lines of the container.

3 Sprinkle the rocket seeds thinly over the surface of the compost in the gaps between the other plants and cover with a light dusting of sieved compost. Water well. Because it is fast maturing, the rocket will have just enough time to get established before the other plants fill out.

Pizza napoletana

Make or buy 2 x 23 cm (9 inch) pizza bases. Brush them with olive oil, then divide among them 400 g (13 oz) canned chopped and drained tomatoes, 1 tablespoon chopped basil, 2 teaspoons fresh oregano, 175 g (6 oz) sliced mozzarella cheese and 4 tablespoons grated Parmesan cheese. Bake in a preheated oven at 220°C (425°F), Gas Mark 7, for 15 minutes, then reduce the heat to 180°C (350°F), Gas Mark 4, for 5 minutes.

Once upon a time

Aubergines are one of the most attractive of all fruiting vegetables. As well as the glossy, deep purple, pink-purple, pale green or white fruits, they bear pretty mauve flowers and handsome felty leaves. They grow well in containers, where it is easy to control watering and make sure they are neither too dry nor too wet.

Ingredients

1 aubergine 'Fairytale F1' plant or a packet of seed

module trays

seed compost

1 packet of dwarf runner bean 'Snow White' seed

biodegradable fibre pots

small plastic plant pots

1 light grey, terrazzo, reconstituted-stone container, 50 x 45 cm (20 x 18 in)

drainage material (see page 16)

multi-purpose soil-less compost

well-rotted manure or garden compost

Method

1 Sow the aubergines in seed compost in module trays in a heated greenhouse or propagating case. Sow the runner beans into individual fibre pots two or three weeks later because they mature more quickly than the aubergines. When they large enough to handle, transplant the aubergines into small plastic pots.

2 Plant out when all danger of frost has passed. Add a layer of drainage material to the container before filling with a good-quality, soil-less compost, adding extra organic matter such as well-rotted manure. Position the aubergine in the centre, surrounded by a ring of beans spaced about 10 cm (4 in) apart.

3 Feed and water regularly. Gently mist over the runner bean flowers occasionally to help them set, and pick the fruits when they are young and tender.

Tomato and aubergine parmigiana

Slice a large aubergine and fry in olive oil until golden-brown. Cut 500 g (1 lb) ripe red plum tomatoes into wedges and arrange them in alternate layers with the aubergine in a shallow ovenproof dish, sprinkling a total of 50 g (2 oz) grated Parmesan cheese between layers. Season to taste and bake in a preheated oven, 190°C (375°F), Gas Mark 5, for 15–20 minutes. Garnish with parsley and serve immediately or at room temperature.

Textural treats

This collection of delicate edible flowers and tasty leaves is complemented by wispy grasses to create a hazy summer display that looks, tastes and smells great. Both the blue-flowered hyssop and pinkish-lavender agastache are attractive to bees, so this is the perfect container to site among your pots of beans, tomatoes and other fruiting plants to ensure their blooms are well pollinated and plenty of fruit sets.

Ingredients

1 wooden crate-style container, 60 x 30 cm (24 x 12 in)

polythene

drainage material (see page 16)

multi-purpose soil-less compost

2 *Agastache pallidiflora* x *neomexicana* 'Lavender Haze' plants (non-edible)

3 *Hyssopus officinalis* (hyssop) plants

5 small *Stipa tenuissima* plants (non-edible)

2 *Carex comans* 'Frosted Curls' plants (non-edible)

3 *Coriandrum sativum* (coriander) plants

2 *Thymus vulgaris* (common thyme) plants

2 *Thymus* Coccineus Group plants

1 packet of *Satureja hortensis* (summer savory) seed

Method

1 Line the crate with polythene and pierce holes in the base. Cover the drainage holes with a layer of drainage material and fill to within 10 cm (4 in) of the top with soil-less compost.

2 While they are still in their pots, lay the plants out on the surface of the compost in a random fashion to check that you are happy with their position before planting. Once planted, fill in with more compost, firm gently and water thoroughly. Sow the summer savory in between to fill in the gaps.

3 Remove faded flowering stems from the agastache and hyssop to encourage new flowering shoots to form and lightly trim the thymes after flowering.

Griddled snapper with red onions and thyme

Cook 2 sliced red onions in a hot griddle until soft. Add a handful of thyme, chopped, and push to the side of the griddle. Cook 4 snapper fillets, each about 175 g (6 oz), for about 4 minutes on each side. Serve with the onions and thyme and a drizzle of olive oil.

Mint medley

Because they are notoriously invasive, mints are best grown in separate pots where their wayward habits can be kept in check. Tall, slender terracotta pots and chimney pots of varying heights show each mint off to perfection.

Ingredients

1 packet of chard or leaf beet 'Bright Lights' seed

module trays

seed compost

small plastic plant pots

2 deep plastic plant pots

drainage material (see page 16)

multi-purpose soil-less compost

1 *Mentha pulegium* (pennyroyal) plant

1 *Mentha spicata* 'Moroccan' (Moroccan mint) plant

1 *Mentha* x *piperita* f. *citrata* (eau-de-cologne mint) plant

1 *Mentha suaveolens* 'Variegata' (pineapple mint) plant

1 *Mentha requienii* (Corsican mint) plant

2 reclaimed chimney pots

1 ribbed terracotta pot, 35 x 60 cm (14 x 24 in)

1 ribbed terracotta pot, 25 x 40 cm (10 x 16 in)

1 plain terracotta pot, 23 x 30 cm (9 x 12 in)

Method

1 Sow the chard or leaf beet in seed compost in module trays and transplant into small individual pots when the seedlings are large enough to handle.

2 Choose two plastic pots – preferably deep ones in which climbing plants are sold – to sit inside the chimney pots.

3 Following the instructions for planting a container on page 16, pot up each of the mints individually using soil-less compost. Pot the pennyroyal and the Moroccan mint into the plastic pots and plant the eau-de-cologne and pineapple mints directly into the tallest and shortest terracotta pots. Plant a single chard in the centre of the medium-sized terracotta pot. Split the Corsican mint into four or five pieces and plant them around the base of the chard.

4 Place the plastic pots inside the chimney pots. When the plants become potbound, lift and split the mints, then replant a smaller portion. Pick leaves regularly for garnishing.

Minted rice with tomato and sprouted beans

Finely chop 6 spring onions and 2 garlic cloves and stir-fry them for 2–3 minutes in 2 tablespoons olive oil. Add 750 g (1½ lb) cooked, cooled Basmati rice and cook for a further 3–4 minutes. Stir in 2 finely chopped ripe plum tomatoes and 250 g (8 oz) mixed sprouted beans (such as aduki, mung, lentil and chickpea sprouts) and cook for 2–3 minutes. Stir in a small handful of mint leaves and serve immediately.

On fire!

There are a huge number of pepper cultivars available. You can use any cultivars you like, depending on your taste. This selection includes 'Nosegay' and 'Paper Lantern', both with hot, red fruits, 'Sweet Orange Baby', a productive, compact, small-fruited sweet pepper, and 'Healthy', a sweet pepper that will ripen in cooler conditions than most cultivars.

Ingredients

1 packet of sweet pepper 'Healthy' seed

1 packet of chilli pepper 'Nosegay' seed

1 packet of chilli pepper 'Paper Lantern' seed

1 packet of sweet pepper 'Sweet Orange Baby' seed

small plastic plant pots

seed compost

1 terracotta bowl planter, 45 x 30 cm (18 x 12 in)

multi-purpose soil-less compost

drainage material (see page 16)

1 *Cymbopogon citratus* (lemon grass) plant

1 plain terracotta pot, 25 x 30 cm (10 x 12 in)

2 *Coriandrum sativum* (coriander) plants

1 *Satureja spicigera* (savory) plant

1 terracotta bowl, 35 x 23 cm (14 x 9 in)

1 bag of reddish-brown pebbles

Method

1 Sow the peppers in seed compost in small pots in mid-spring, two seeds to a pot, and put them in a heated propagating case or greenhouse. Keep well watered but not too wet. If both seeds in a pot germinate, pull out the weaker seedling.

2 When they are growing strongly and all danger of frost has passed, plant the strongest pepper plants – one of each cultivar – into a wide terracotta bowl filled with soil-less compost. Follow the instructions for planting a container on page 16. Firm and water well. Pot up the lemon grass in the tall terracotta pot and the coriander and savory together in the smaller terracotta bowl in the same way. Mulch the containers with a layer of reddish-brown pebbles.

3 Group the pots together in as sunny a position as possible to encourage a good crop of peppers and to help them ripen up as quickly as possible.

Spicy tomato salsa

Finely chop 1 small red onion and 1 garlic clove. Skin, deseed and chop 500 g (1 lb) sweet ripe tomatoes. Deseed and finely chop 2 red chillies. Put the onion, garlic, tomatoes and chillies into a bowl and add 3 tablespoons of finely chopped coriander, 1 tablespoon of lime juice and 3 tablespoons of olive oil. Season with a pinch of sugar and salt and mix lightly. Cover and chill for 30–60 minutes.

Potato paradise

Potatoes are easy to grow in containers, and even a pot 30 cm (12 in) across with a single tuber will give a good crop. Although you can grow any potato in a pot, the best for containers are salad types, such as 'Mimi', which will give a high yield of small, well-formed, pink-skinned tubers.

Ingredients

9 tubers of salad potato 'Mimi'

1 slatted wooden container, 60 x 60 cm (24 x 24 in)

multi-purpose soil-less compost

1 *Hyssopus officinalis* (hyssop) plant

1 *Thymus* 'Peter Davis' (thyme) plant

1 *Mentha spicata* var. *crispa* 'Moroccan' (Moroccan mint) plant

1 *Mentha* x *piperita* (peppermint) plant

1 *Petroselinum crispum* (parsley) plant

4 plain terracotta pots of varying sizes

horticultural grit

high-potash fertilizer

Method

1 Plant the potatoes, evenly spaced, towards the bottom of a large container that is at least 35 cm (14 in) deep and cover with a 5–10 cm (2–4 in) layer of multi-purpose soil-less compost.

2 When the potatoes are growing strongly, fill around them with more compost in the same way as you would earth them up in the garden. Do this two or three times, as they grow.

3 Team up the hyssop and thyme in the largest pot and plant the mints and parsley individually. Following the instructions for planting a container on page 16, plant them in a soil-less compost with added grit to improve the drainage.

4 Do not overwater the potatoes because potato blight is more prevalent in wet conditions. Feed with a high-potash fertilizer and harvest when the plants have flowered and are beginning to die down.

Griddled warm new potatoes with fresh mint dressing

Cut 750 g (1½ lb) small new potatoes in half lengthways and cook them on a hot griddle for 6 minutes on each side. Meanwhile, mix the finely grated rind and juice of 2 limes with 8 tablespoons grapeseed oil. Season to taste and add 2 tablespoons chopped fresh mint. Toss the potatoes in the dressing and serve garnished with extra mint leaves.

Peas and beans

As they require the same fertile, moisture-retentive but free-draining conditions, peas and beans make good bedfellows. 'Borlotto Firetongue' is a dwarf French bean with pods that are streaked pinkish-red and green, while 'Ferrari', with its plain green pods, is one of the smallest of all French beans. Centre stage is given to a group of sprawling asparagus peas. Their chief attraction are the small red flowers, which are followed by small, edible winged pods.

Ingredients

1 packet of dwarf French bean 'Ferrari' seed

1 packet of dwarf French bean 'Borlotto Firetongue' (or a similar cultivar) seed

1 packet of *Lotus tetragonolobus* (asparagus pea or winged pea) seed

biodegradable fibre pots

seed compost

1 wooden windowbox, 70 x 25 cm (28 x 10 in)

drainage material (see page 16)

multi-purpose soil-less compost

6–8 twiggy sticks, about 45 cm (18 in) long, for support

Method

1 Sow the beans and asparagus peas in seed compost in individual fibre pots to prevent root disturbance when planting out, and put them in a heated propagating case or greenhouse to germinate.

2 Make sure there are plenty of drainage holes in the base of the windowbox, and drill extra ones if necessary. Put a layer of drainage material in the base and fill with multi-purpose soil-less compost.

3 Plant seven or eight evenly spaced 'Borlotto Firetongue' plants in a row towards the back of the trough. In the front row plant three asparagus peas spaced 20 cm (8 in) apart, one dead centre, the others on either side. Finally, plant two evenly spaced 'Ferrari' beans between the asparagus peas. Push in a few twiggy sticks to help support the plants as they grow.

French bean and tomato salad

Cook 250 g (8 oz) French beans and cut 250 g (8 oz) mixed red and yellow baby tomatoes in half. Put the beans and tomatoes in a bowl and mix in a handful of chopped mint, 1 chopped garlic clove, 4 tablespoons olive oil and 1 tablespoon balsamic vinegar. Season to taste and serve warm or cold.

Pretty in purple

Handsome-looking aubergine 'Baby Rosanna F1' bears heavy crops of small but tasty, purple fruits. An underplanting of evergreen tri-coloured sage, with pink, cream and green, aromatic leaves that can be used in the same way as common sage, is the perfect foil. Interspersed with the sage are pink- and blue-flowered annual clary, the leaves of which can also be used for flavouring.

Ingredients

1 aubergine 'Baby Rosanna F1' plant (or similar small-fruited variety) or a packet of seed

module trays

seed compost

1 packet of *Salvia viridis* (annual clary) seed

small plastic plant pots

1 grey slate container, 38 x 38 cm (15 x 15 in)

drainage material (see page 16)

multi-purpose soil-less compost

4 *Salvia officinalis* 'Tricolor' (tri-coloured sage) plants

high-potash fertilizer

Method

1 Sow the aubergine seed in seed compost in module trays in a heated propagating case in mid-spring, and the faster maturing annual clary in a cold frame a few weeks later. Pot the aubergine seedlings on into small individual pots once they are large enough.

2 Add a layer of drainage material to the base of the slate container before filling with soil-less compost. When all danger of frost has passed, plant the strongest, bushiest aubergine (or a bought plant) in the centre. Position one of the tri-coloured sages at each corner and infill with annual clary seedlings.

3 Pinch out the clary to make the plants bush out and trim off the flowering stems as they fade. Feed the aubergine regularly with a high-potash fertilizer.

Aubergine and yogurt wraps

Thinly slice an aubergine and cook in 4 tablespoons olive oil for about 10 minutes. Chop a handful of mint and of parsley and mix with 2 tablespoons chopped chives and 1 green chilli, deseeded and sliced. Add to 200 ml (7 fl oz) Greek yogurt and 2 tablespoons mayonnaise. Arrange aubergine slices on 2 large tortillas and spread the yogurt mixture over the top. Top with cucumber slices and sprinkle with paprika.

Roots and shoots

Kohl rabi is a fast-growing, easy plant, and the turnip-like stems should be harvested when they reach the size of golf balls. Being shallow-rooted, drought-tolerant creeping thymes offer little competition for water and nutrients, while their tiny, green-and-golden-yellow leaves provide the perfect accompaniment to the dramatic-looking kohl rabi.

Ingredients

1 packet of mixed purple and white kohl rabi seed

module trays

seed compost

small plastic plant pots

1 terracotta trough, 50 x 30 cm (20 x 12 in)

drainage material (see page 16)

multi-purpose soil-less compost

6 *Thymus serpyllum* 'Goldstream' plants (non-edible)

Method

1 Sow the kohl rabi in seed compost in module trays, two or three seeds to a compartment. When they have germinated, remove the weakest seedlings to leave the strongest to develop. Once they are large enough, pot up the best seedlings into small individual pots.

2 Add a layer of drainage material to the base of the trough and fill with soil-less compost. Select an even number of purple and white plants – the purple plants have darker stems – and plant them 10 cm (4 in) apart, alternately through the centre of the trough.

3 Carefully split each of the thyme plants into two equal-sized pieces and plant them around the kohl rabi. They will quickly knit together to form a carpet of green-and-golden-yellow leaves.

Minty carrot and kohl rabi salad

Mix together 250 g (8 oz) thinly sliced carrots, 150 g (5 oz) thinly sliced kohl rabi, 200 ml (7 fl oz) water, 65 ml (2½ fl oz) white wine vinegar, 1 tablespoon soft brown sugar and ½ teaspoon sea salt. Transfer to the refrigerator for about 1 hour, stirring occasionally. Drain the vegetables and rinse in water. Mix in 2 tablespoons chopped mint leaves and 2 tablespoons chopped fresh coriander leaves, and serve.

Red devil

This simple but highly effective planting will carry your edible container display well into the cooler, darker months, providing an invaluable supply of leaves for winter use. Radicchio forms tight, succulent heads of glossy leaves that, in most varieties, are green at first, becoming increasingly flushed with red or bronze as they mature. Rising up in the centre of the container, ruby-red chard creates a colourful display of glowing red stems and red-veined leaves.

Ingredients

1 packet of radicchio seed

1 packet of red-stemmed chard or leaf beet seed

seed compost

module trays

1 old galvanized bucket, 40 x 25 cm (16 x 10 in)

drainage material (see page 16)

multi-purpose soil-less compost

Method

1 Sow the radicchio and chard in seed compost in module trays. Place them in a cold frame to germinate and keep them moist but not overwet.

2 Drill holes in the base of the bucket for drainage, wearing goggles to protect your eyes from metal splinters. Place a layer of drainage material in the base and fill with multi-purpose soil-less compost.

3 Once they are large enough to handle, plant the seedlings into the container, positioning a single chard plant in the centre surrounded by four radicchio. Keep well watered, and place in a cool but light spot to prevent the plants bolting prematurely.

Chard and chickpea tortilla

Heat 4 tablespoons of olive oil in a large frying pan. Add 1 chopped onion, 4 crushed cloves of garlic and ½ teaspoon dried chilli flakes and fry gently for 10 minutes. Stir in 500 g (1 lb) shredded chard and 400 g (13 oz) canned chickpeas and cook gently for 5 minutes. Beat 6 eggs in a bowl and season with salt and pepper. Stir in the chard mixture. Wipe out the frying pan and add 4 tablespoons of olive oil. Pour in the chard/egg mixture and cook over a low heat for 10 minutes. Slide the tortilla onto a plate, invert the pan over the plate and flip it back in the pan. Return to the heat for 5 minutes until the tortilla is cooked through.

Pepper pot

This fiery concoction combines edible fruits, flowers and leaves to create a sizzling display that will last all summer. The tall pot raises the plants nearer to eye level, where this simple but productive planting can be more readily appreciated. The sweet pepper 'Redskin F1' is a compact cultivar, producing good-sized fruits that turn from green to glistening red as they ripen.

Ingredients

1 pale grey, tapered, terrazzo, reconstituted-stone container, 35 x 70 cm (14 x 28 in)

drainage material (see page 16)

multi-purpose soil-less compost

1 sweet pepper 'Redskin F1' plant

4 *Tropaeolum majus* 'Red Wonder' (nasturtium) plants

4 *Calibrachoa* Million Bells Crackling Fire ('Sunbelfire') plants (non-edible)

high-potash fertilizer

Method

1 Make sure there are plenty of drainage holes in the base of the stone container and, if not, carefully drill a few extra ones. Add a layer of drainage material in the base and fill with soil-less compost.

2 Once all danger of frost has passed, plant the pepper in the centre of the pot, surrounded by nasturtiums and then the calibrachoas to trail over and soften the edges of the pot.

3 Harvest the peppers as they ripen and apply with a high-potash fertilizer to encourage successive crops. Deadhead the nasturtiums to keep them flowering well and keep an eye out for aphids.

Chicken and sweet pepper kebabs

Cut 8 boneless chicken thighs, 1 onion, 1 red pepper and 1 green pepper into chunks of the same size. Mix together 150 ml (¼ pint) natural yogurt, 2 tablespoons olive oil, 2 crushed garlic cloves, 2 tablespoons chopped fresh coriander and 2 tablespoons ground cumin. Stir in the chicken and refrigerate for about an hour. Thread the chicken, onion and peppers on to skewers and cook under a hot grill for 20 minutes, turning frequently. Serve immediately.

Purple and bronze

Bronze fennel and purplish-pink-flowered marjoram alternate at the back of this large, deep, galvanized steel trough to create a tactile backdrop to the bronze-leaved para cress. The graceful fennel, with its aniseed-flavoured foliage, can grow to 1.5 m (5 ft) tall, and it will self-sow if you allow the flowerheads to set seed. The marjoram has leaves that are purple-flushed in spring. The yellow-bronze blooms of the para cress echo the colouring of the fennel leaves.

Ingredients

1 packet of *Acmella oleracea* (para cress) seed

module trays

seed compost

small plastic pots

1 galvanized steel trough, 60 x 30 cm (24 x 12 in)

drainage material (see page 16)

multi-purpose soil-less compost

3 *Foeniculum vulgare* 'Purpureum' (bronze fennel) plants

2 *Origanum laevigatum* 'Herrenhausen' (marjoram) plants

Method

1 Sow the para cress in seed compost in module trays in spring. Transfer them to a cold frame to germinate and transplant the seedlings into individual pots when they are large enough to handle.

2 Add a layer of drainage material to the base of the trough and fill with soil-less compost. Plant three strong, bushy fennel plants and two equal-sized marjorams in a row along the back of the container, alternating between fennel and marjoram.

3 Plant out the para cress plantlets at the front of the trough and water well. Pinch out the young leaves and shoot tips and use them in salads.

Tabbouleh and fennel salad

Prepare 250 g (8 oz) bulgar wheat according to the directions on the packet, then tip it into a large bowl. Stir in 1 finely sliced fennel bulb, 1 finely sliced red onion, 5 tablespoons chopped mint, 5 tablespoons chopped parsley, 2 tablespoons crushed fennel seeds, 2 tablespoons olive oil, the finely grated rind of 2 lemons and the juice of 1 lemon. Season with salt and pepper, allow to stand for 30 minutes, then serve, with more lemon juice if wished.

A taste of the Mediterranean

Taking centre stage in this terracotta pot is a mini-standard olive tree that, given plenty of warmth, will produce a small but invaluable crop. Its grey-green leaves are an attractive contrast to an underplanting of cut-leaf rocket and bright green-leaved sweet basil.

Ingredients

1 plain terracotta pot, 42 x 35 cm (17 x 14 in)

drainage material (see page 16)

soil-based potting compost

multi-purpose soil-less compost

1 mini-standard *Olea europea* (olive) plant

3 *Ocimum basilicum* (sweet basil) plants

1 packet of cut-leaf *Eruca vesicaria* (annual rocket) seed

Method

1 Add a layer of drainage material to the base of the pot and quarter-fill the container with a half-and-half mix of soil-based and soil-less compost. Set the olive in the centre, adding or removing compost as necessary until the top of the rootball rests 5 cm (2 in) below the rim of the pot.

2 Fill in with more compost and plant three sweet basils evenly spaced around the olive. Sow the rocket seed direct on to the surface of the compost between the basils, and cover lightly.

3 Ensure that the compost is kept moist but not overwet, and position the pot in a sunny but sheltered spot. The basils and rocket are annuals; but, if it is overwintered in a frost-free place, the olive can be kept from year to year and repotted into a larger container as necessary.

Panzanella

Tear 4 slices of ciabatta bread into pieces and put them in a bowl. Cut 4 tomatoes, ½ cucumber and 1 red onion into similar sized pieces. Mix in a handful of chopped flat leaf parsley and 1 tablespoon chopped black olives. In a separate bowl mix 4 tablespoons olive oil, 1–2 tablespoons wine vinegar and 2 tablespoons lemon juice. Pour the dressing over the salad and leave to stand for 1 hour at room temperature before serving.

Kale and cabbage

Cabbage does well in fairly cool conditions, and red cabbage is usually grown for picking in late summer to autumn. Here the cultivar 'Red Jewel' is set in a triangle in a sea of 'Dwarf Green Curled' kale in a large rough-cast, mellow stone-effect concrete container.

Ingredients

1 packet of red cabbage (such as 'Red Jewel') seed

1 packet of compact-growing green kale (such as 'Dwarf Green Curled') seed

module trays

seed compost

1 stone-effect concrete container, 40 x 35 cm (16 x 14 in)

drainage material (see page 16)

multi-purpose soil-less compost

Method

1 Sow the cabbage and kale in seed compost in module trays in early spring for a summer crop or in early summer for winter harvesting. Transfer to a cold frame to germinate.

2 When they are large enough to handle and growing strongly, plant them out in a large container, following the instructions for planting a container on page 16. Set three cabbages in a triangle formation with the kale filling in the gaps between. Plant the kale fairly close together so that it can be thinned out, and the thinnings used as baby leaves in salads.

3 Leave some kale to mature for use later in the season and to keep the display looking good. Keep an eye out for caterpillars at all times, because they can eat their way through a crop in hours. Small numbers can be simply picked off.

Cabbage with pancetta

Fry 1 sliced onion, 1 crushed garlic clove, 1 diced red chilli and 125 g (4 oz) diced pancetta in 1 tablespoon olive oil until soft. Cut 1 head cabbage in half lengthways, discard the hard central stem and roughly chop the leaves. Add to the onion mixture with 75 ml (3 fl oz) chicken stock, stir well, season with salt and pepper and cook for 4 minutes. Sprinkle 75 g (3 oz) grated Parmesan cheese and 2 tablespoons chopped parsley over the top and serve immediately.

On the bay

This aromatic gathering of evergreen culinary herbs will provide a year-round display. When planted in a wooden Versailles-style container and clipped as a topiary lollipop, sweet bay forms a classic centrepiece. The accompanying terracotta pots are filled with violet-flowered rosemary and scented-leaf pelargoniums.

Ingredients

1 quarter-standard *Laurus nobilis* (sweet bay), clipped as a lollipop

1 square wooden Versailles planter, 40 x 40 cm (16 x 16 in)

drainage material (see page 16)

multi-purpose soil-less compost

soil-based potting compost

1 *Rosmarinus officinalis* 'Severn Sea' (rosemary) plant

1 *Pelargonium* 'Shottesham Pet' plant

1 *Pelargonium* 'Lady Plymouth' plant

3 plain round terracotta pots in an assortment of sizes, 15–25 cm (6–10 in) across

1 bag black, river-polished pebbles

Method

1 Choose a straight-stemmed, quarter-standard bay lollipop from your local nursery or garden centre. It may be expensive but it will last for years.

2 Add a layer of drainage material to the base of a Versailles planter and plant the bay in the container using a half-and-half mix of soil-less and soil-based composts. In the same way, pot up the rosemary and pelargoniums in individual terracotta pots, this time using soil-less compost.

3 Group the pots together to create a pleasing composition and mulch the surface of the Versailles planter with black polished pebbles for a contemporary look. Trim the bay two or three times through the season to keep its shape, and trim the rosemary after flowering.

Sea bass in cartoccio

Mix the rind from 2 oranges with 2 tablespoons olive oil. Slice the orange flesh and arrange half the slices on 4 large rectangles of greaseproof paper. Place the sea bass on the slices and insert a bay leaf in each cavity and one on top. Put the remaining orange slices on top. Drizzle over the oil and rind mixture, fold the paper around the fish and bake in a preheated oven, at 190°C (375°F), Gas Mark 5, for 20 minutes.

Fireball

These dwarf chilli peppers are perfectly happy growing in a large hanging basket, where they create an explosion of fiery colour in mid- to late summer. The centrepiece is the reliable 'Apache', which is crowded with small, tapered fruits that turn from rich, glossy green to startling bright red.

Ingredients

1 chilli pepper 'Apache' plant

4 chilli pepper 'Prairie Fire' plants (or similar dwarf cultivar) or 1 packet of seed

small plastic plant pots (optional)

seed compost (optional)

1 willow or rattan hanging basket, 35 cm (14 in) across

plastic liner

multi-purpose soil-less compost

4 *Lotus maculatus* x *berhelottii* (parrot's beak) plants (non-edible)

Method

1 If you are growing them from seed, sow the chilli peppers in seed compost in small individual pots in mid-spring, two seeds to a pot, and put them in a heated propagating case or greenhouse. Keep well watered but not too wet. If both seeds germinate, pull out the weaker seedling.

2 Choose a large willow or rattan closed-sided hanging basket and pierce several holes in the base of the plastic lining. Fill with a good-quality, soil-less compost.

3 Plant the taller chilli in the centre with the plants of the other cultivar spaced evenly around it. Fill in the gaps with the parrot's beak plants. Fill around the plants with more compost and water well.

4 Stand the basket on a pot and leave it in a greenhouse or conservatory to settle for a few weeks before hanging in its final position, once all danger of frost has passed. Choose as sunny a position as possible to make sure that you get a good crop of colourful fruits.

Chicken with chilli jam

Put 125 g (4 oz) deseeded and chopped chillies, 1 clove of crushed garlic, 1 chopped onion, 5 cm (2 inch) chopped root ginger, 125 ml (4 fl oz) vinegar and 500 g (1 lb) sugar into a saucepan, bring to the boil, then simmer for 15 minutes until it becomes thick and jam-like. Meanwhile heat a griddle pan. Cook 4 chicken breasts on the griddle, skin-side down, for 10 minutes on each. Serve the chicken with the chilli jam over the top.

Green garnish

This delightful container presents a selection of the most useful garnishing herbs. Parsley is a hardy biennial forming mounds of fresh-green leaves, while coriander is grown both for its leaves and the seeds that follow the tiny white summer flowers. Rocket is a favourite Mediterranean garnish and the long flat leaves of Chinese chives add a garlic flavouring when sprinkled over dishes.

Ingredients

1 metal bucket, 35 x 20 cm (14 x 8 in)

drainage material (see page 16)

multi-purpose soil-less compost

2 *Eruca vesicaria* (annual rocket) plants

1 *Petroselinum crispum* (parsley) plant

1 *Allium tuberosum* (garlic or Chinese chives) plant

1 *Coriandrum sativum* (coriander) plant

Method

1 Turn the bucket upside down and drill holes in the base for drainage. Wear goggles to protect your eyes from metal splinters.

2 Add a layer of drainage material in the base and half-fill the container with multi-purpose soil-less compost.

3 Position the plants, placing the shorter-growing rocket towards the front, with the medium-growing parsley and chives to each side and the tall-growing coriander at the back. Fill in with more compost, firm gently and water well. Pick regularly to encourage the production of abundant new leaves.

Spiced couscous salad

Combine 200 ml (7 fl oz) vegetable stock with 200 ml (7 fl oz) orange juice, 1 teaspoon ground cinnamon, ½ teaspoon ground coriander and ½ teaspoon salt in a saucepan. Bring it to the boil, stir in 250 g (8 oz) couscous and remove the pan from the heat. Cover and leave to stand for 10 minutes. Combine 75 g (3 oz) raisins, ½ bunch of roughly chopped parsley, ½ bunch of roughly chopped mint, 1 crushed garlic clove and 4 tablespoons of oil in a bowl. Stir in the soaked couscous, and season with salt and pepper.

Lots of leaves

Chinese cabbage 'Santo' and radicchio 'Palla Rossa Red Devil' give a display of shaped and textured leaves that is shown off to perfection in an earthenware container. Both can be left to mature and be harvested as whole heads, or you can pick a few leaves at a time. In a container, however, they are best planted more closely than in the ground, harvested when young and replaced with another crop.

Ingredients

1 packet of Chinese cabbage (such as 'Santo') seed

1 packet of radicchio (such as 'Palla Rossa Red Devil') seed

module trays (optional)

seed compost (optional)

1 earthenware container, 40 x 45 cm (16 x 18 in)

drainage material (see page 16)

multi-purpose soil-less compost

Method

1 Sow the Chinese cabbage and radicchio in seed compost in module trays in summer, for harvesting in autumn. Alternatively, sow directly into the container and thin out as necessary.

2 Module-raised plants should be carefully transplanted as soon as they are strong enough to handle. Following the instructions for planting a container on page 16, plant out into multi-purpose soil-less compost, positioning a block of Chinese cabbage in the centre with a ring of radicchio to surround them.

3 Position the container in a sunny but not excessively hot position, because too much heat may cause them to bolt. Keep the plants well watered and watch out for snails and caterpillars.

Radicchio with pears and Stilton

Cut 4 pears into quarters and arrange them in a single layer on a sheet of kitchen foil, turning up the edges. Mix together the juice and grated rind of 2 oranges and 4 tablespoons clear honey. Pour the mixture over the pears and press the edges of the foil together to seal. Transfer to a frying pan and cook over a moderate heat for 15–20 minutes. Meanwhile, cut 4 small heads of radicchio into quarters, brush with walnut oil and cook under a grill for 2–3 minutes on each side. Arrange the pears and juices on plates with the raddicchio, and crumble over 125 g (4 oz) Stilton cheese.

Desserts

An apple a day **112**

Blueberry surprise **114**

Lemon zest **116**

Strawberry ball **118**

Gorgeous grapes **120**

Currant affair **122**

Mellow yellow **124**

Tea pot **126**

Pop and go **128**

Gorgeous gourds **130**

Pear delight **132**

Passion fashion **134**

An apple a day

*begonias
only*

Apples grown on a dwarfing or very dwarfing rootstock will grow to about 1.8 m (6 ft) tall after ten years or so and are ideal for containers. When grown in a generous-sized container and fed with a high-potash fertilizer from blossom time until shortly before harvest, they will produce a worthwhile crop.

Ingredients

1 wooden half-barrel, 70 x 40 cm (28 x 16 in)

drainage material (see page 16)

soil-based potting compost

multi-purpose soil-less compost

1 apple 'Cox's Orange Pippin Self Fertile' on a very dwarfing rootstock (such as M27)

wooden stake (optional)

8 *Begonia* 'Non Stop Appleblossom' plants

Method

1 Choose a large wooden half-barrel and site it in a sunny but sheltered position. Once planted, the container will be heavy and difficult to move, so make sure it is in exactly the right spot and plant up in situ.

2 Put a layer of drainage material in the base of the container and part fill with a half-and-half mix of soil-less and soil-based composts. Position the apple tree in the centre and fill in with more compost, firming gently. The roots of dwarf apple trees are generally fairly shallow and compact so it may be necessary to provide a wooden stake for support in exposed locations.

3 Plant a ring of begonias around the tree. In future years the roots of the apple will prevent replanting at its base, but you could sprinkle seeds of hardy annuals on the surface of the compost, because they will offer minimal root competition.

Apple tartlets with passionfruit cream

Cut 4 circles, 5 mm (¼ inch) thick, from 250 g (8 oz) puff pastry. Put them on an oiled baking sheet and prick. Peel, core and slice 3 apples and arrange the slices on the pastry. Pour over 25 g (1 oz) melted butter and sprinkle with 100 g (3½ oz) sugar. Cook in a preheated oven, at 200°C (400°F), Gas Mark 6, for 15–20 minutes. Mix the flesh of 2–3 passionfruit with 125 ml (4 fl oz) cream and serve with the apple tartlets.

Blueberry surprise

Not only do blueberries produce heavy crops of delicious fruits, they also offer great visual appeal. In spring there are sweetly scented, creamy-white flowers; in summer the dark blue fruits appear; and in autumn the leaves turn shades of gold and red before they fall. Cranberries enjoy the same acidic soil conditions and, being low and trailing, are excellent for underplanting blueberries.

Ingredients

drainage material (see page 16)

1 dark blue-green, round, glazed container, 50 x 45 cm (20 x 18 in)

ericaceous (lime-free) compost

1 *Vaccinium corymbosum* 'Elliott' (blueberry) plant

3 *Vaccinium oxycoccos* (cranberry) plants

1 bag of cocoa shells

ericaceous fertilizer

Method

1 Put a layer of drainage material in the base of the container and half-fill it with ericaceous compost.

2 Sit the blueberry in the centre of the pot so that the top of its rootball rests 8 cm (3 in) below the rim of the pot. Evenly space the cranberries around it, allowing their stems to trail over the edges.

3 Fill in with more compost and firm gently. Water well and mulch with cocoa shells. Water with rainwater, if possible, and feed occasionally during the growing season with an ericaceous fertilizer. If necessary, trim the cranberries during spring.

Blueberry and mascarpone gratin

Arrange 50 g (2 oz) blueberries on 2 heatproof plates or in gratin dishes. Beat together 50 g (2 oz) mascarpone cheese, 1 egg yolk, 25 g (1 oz) caster sugar and 1 tablespoon Amaretto di Saronno. Spoon the smooth mixture over the fruits and cook under a hot grill for about 3 minutes, until the sauce is caramelized and the fruits have softened.

Lemon zest

When underplanted with lemon-scented pelargoniums and accompanied by a billowing pot of tangerine sage, this fruiting lemon tree creates not only a visual feast but also a treat for the nostrils and tastebuds. The variety 'Meyer' is a compact hybrid that bears fragrant white flowers from spring to summer, followed by small but well-formed fruits. Keep it outdoors on a sunny patio in summer but move to a frost-free position in winter.

Ingredients

1 *Citrus x meyeri* 'Meyer' (lemon) plant

1 cylindrical terracotta pot, 40 x 40 cm (16 x 16 in)

drainage material (see page 16)

soil-based potting compost

multi-purpose soil-less compost

3 *Pelargonium* 'Lady Plymouth' plants

1 *Salvia elegans* (tangerine sage) plant (non-edible)

1 cylindrical terracotta pot, 30 x 30 cm (12 x 12 in)

balanced fertilizer

Method

1 Pot the lemon into the larger container, placing a layer of drainage material in the base. Use a half-and-half mix of soil-based and soil-less composts.

2 Position the three pelargoniums evenly around the lemon, setting them towards the edge of the pot. Infill with more compost.

3 Plant the tangerine sage in the smaller pot. Remove faded flowers from the pelargoniums. While it is in growth, feed the lemon tree with a balanced fertilizer every two to three weeks..

Lemon and cinnamon pancakes

Sift 125 g (4 oz) plain flour, ½ teaspoon ground cinnamon, a pinch of salt and 1 teaspoon lemon rind into a bowl. Gradually beat in 1 egg, 300 ml (½ pint) milk and 15 g (½ oz) melted butter to make a smooth batter. Brush a small pan with a little oil and fry a ladleful of batter for a minute before flipping it over. Continue until all the batter is used. Serve 2–3 crêpes per person.

Strawberry ball

Strawberries are among the easiest of all fruits to grow in containers. Fed and watered well, they will thrive even within the confines of a hanging basket. Indeed, growing strawberries so that they are raised off the ground makes them less accessible to slugs and snails. Choose a perpetual or ever-bearing cultivar, such as 'Elan', which will provide a supply of succulent fruits from early summer right through to autumn.

Ingredients

1 open-sided, wire or plastic hanging basket, 35 cm (14 in) across

1 hanging basket liner suitable for a 35 cm (14 in) basket

multi-purpose soil-less compost

9 perpetual-fruiting strawberry plants, such as 'Elan'

netting (optional)

high-potash fertilizer

Method

1 Line the hanging basket with a flexible liner such as coconut fibre matting or a synthetic moss substitute and fill with soil-less compost to a third of its depth.

2 Choose nine healthy, virus-free, young strawberry plants and carefully push four of them through the side of the basket, about halfway up, making holes in the lining to do so.

3 Add more compost and plant up the top of the basket. Use one plant in the centre and the remaining four spaced alternately to those in the layer below. If necessary, cover with netting to protect from birds. Water regularly and feed weekly with a high-potash fertilizer.

Fruit skewers with chocolate sauce

Break 50 g (2 oz) milk chocolate and 75 g (3 oz) plain chocolate into small pieces and put them in a pan with 75 ml (3 fl oz) milk and 2 teaspoons golden syrup. Heat gently, stirring, until melted and smooth. Meanwhile, cut 1 red-skinned apple, 2 peaches, 1 kiwi fruit, 250 g (8 oz) strawberries and 1 banana into bite-size chunks to dunk into the warm sauce.

Gorgeous grapes

Confined in a container, where they will grow less vigorously, grape vines make handsome specimens, their leaves turning to fiery shades in autumn. This gnarled-stemmed standard grape is underplanted with ajugas, the coloured leaves of which, although edible, are unpalatable so are used here purely for their ornamental appeal.

Ingredients

1 earthenware container, 50 x 40 cm (20 x 16 in)

drainage material (see page 16)

1 standard grape vine (*Vitis vinifera* 'New York Muscat' or similar dessert variety)

soil-based potting compost

multi-purpose soil-less compost

horticultural grit

3 *Ajuga reptans* 'Braunherz' plants (or similar bronze-leaved cultivar)

3 *Ajuga reptans* 'Golden Beauty' plants (or similar variegated-leaf cultivar)

Method

1 Position the container in a warm, sunny spot, preferably near a wall where the grapes can bask in the reflected heat. Vines resent being overwet, so place a deep layer of drainage material in the base of the container. Pot up the vine using a half-and-half mix of soil-based and soil-less composts with added grit to keep the compost open and free draining.

2 Plant the ajugas around the edge of the pot so that they will both creep inwards as well as tumble over the edge.

3 During the growing season feed the vine every two to three weeks. Spray for mildew and other fungal problems as necessary, and carry out restrictive pruning in summer and winter to create a compact framework and plenty of fruiting spurs.

Lemon posset with frosted grapes

Gently heat 300 ml (½ pint) double cream, 75 g (3 oz) caster sugar and the grated rind of ½ lemon until the sugar has dissolved. Simmer until the cream bubbles around the edges. Take off the heat and add 4 tablespoons lemon juice. Pour into 4 dishes and chill for 4 hours. Dip 150 g (5 oz) grapes, snipped into small bunches, in beaten egg white and then coat in caster sugar. Serve on top of the chilled posset.

Currant affair

Redcurrants are among the most ornamental of all soft fruits, the glistening red fruits hanging conspicuously in long strings. For a truly striking display, we've teamed compact redcurrant 'Stanza' with an underplanting of powerfully perfumed Dianthus 'Raspberry Sundae', the petals of which are also edible. 'Stanza' is an early to mid-season cultivar, which crops over a long period, producing small, slightly acidic fruits.

Ingredients

1 charcoal-grey, square, terrazzo, reconstituted-stone container, 40 x 40 cm (16 x 16 in)

drainage material (see page 16)

soil-based potting compost

multi-purpose soil-less compost

1 redcurrant 'Stanza' plant or similar compact variety

8 *Dianthus* 'Raspberry Sundae' (garden pink) plants (or similar dwarf variety)

Method

1 Put a layer of drainage material in the base of the container and fill with a half-and-half mix of soil-based and soil-less composts. Plant the redcurrant in the centre of the container and position the dianthus symmetrically around it.

2 Remove faded flowers from the dianthus regularly and, after flowering and fruiting, move the pot to a less prominent position. Transplant the dianthus to a sunny, well-drained spot in the garden.

3 Cut back all sideshoots produced by the redcurrant during the current season to four or five leaves from their base in early summer. Feed regularly while the fruit is developing.

Summer berry sorbet

Blend 250 g (8 oz) mixed summer berries, 75 ml (3 fl oz) spiced berry cordial, 2 tablespoons Kirsch and 1 tablespoon lime juice in a food processor, taking care not to make it too smooth. Transfer the sorbet to a chilled plastic container and freeze for at least 25 minutes. Spoon into bowls and serve.

Mellow yellow

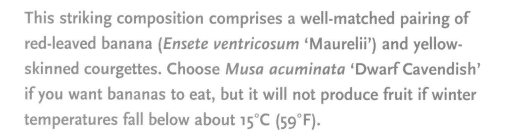

This striking composition comprises a well-matched pairing of red-leaved banana (*Ensete ventricosum* 'Maurelii') and yellow-skinned courgettes. Choose *Musa acuminata* 'Dwarf Cavendish' if you want bananas to eat, but it will not produce fruit if winter temperatures fall below about 15°C (59°F).

Ingredients

1 packet courgette 'Orelia F1' seed

biodegradable fibre pots

seed compost

1 earthenware container, 40 x 45 cm (16 x 18 in)

drainage material (see page 16)

multi-purpose soil-less compost

1 banana plant (use *Musa acuminata* 'Dwarf Cavendish' if you want edible fruit or *Ensete ventricosum* 'Maurelii' for purely ornamental appearance)

1 bag of assorted pebbles

balanced fertilizer

Method

1 In mid- to late spring sow the courgette seed in seed compost in individual fibre pots and place them in a heated greenhouse or propagating case. Once they are large enough, move to a sheltered cold frame and plant out once all danger of frost has passed.

2 Position the container where it will be sheltered from strong winds. Make sure that there are plenty of drainage holes in the base of the container and add a layer of drainage material. Fill with a good-quality, soil-less compost and plant the banana at the back of the container.

3 Plant a strong, healthy courgette plant in front of the banana and cover the exposed compost with a layer of pebbles. Feed every couple of weeks with a balanced fertilizer during the growing season, and remove damaged and yellowing lower leaves as necessary. Harvest the courgettes when they are about 10–15 cm (4–6 in) long.

Carrot and courgette cupcakes

Line a 12-section bun tray. Put 125 g (4 oz) butter, 125 g (4 oz) sugar, 150 g (5 oz) flour, 1 teaspoon baking powder, 1 teaspoon mixed spice, 75 g (3 oz) ground almonds, 2 eggs and the rind of half an orange in a mixing bowl and beat until light and creamy. Add 75 g (3 oz) grated carrots, 75 g (3 oz) grated courgette, 50 g (2 oz) sultanas and stir in until combined. Divide the mixture evenly among the cake cases. Bake in a preheated oven, at 180°C (350°F), Gas Mark 4, for 25 minutes until risen. Leave to cool in the bun tray.

Tea pot

This dark green, glazed, earthenware pot contains a trio of handsome herbs that provide contrasting textures and aromas. Peppermint is one the most strongly aromatic of all and, in common with lemon verbena, its leaves can be used to make herbal teas and tisanes. Nestling at the front of the pot is Roman camomile, which forms a mat of fresh leaves, topped in summer by yellow-centred, white daisy-like flowers that can be used to make a soothing camomile tea.

Ingredients

drainage material (see page 16)

1 octagonal, glazed pot, 35 x 35 cm (14 x 14 in)

multi-purpose soil-less compost

1 *Aloysia triphylla* (lemon verbena) plant

1 *Mentha* x *piperita* (peppermint) plant

2 plastic pots, each 2 litre (3½ pint)

2 *Chamaemelum nobile* (Roman camomile) plants

Method

1 Place a layer of drainage material in the base of the octagonal pot and half-fill with good-quality soil-less compost.

2 Pot up the lemon verbena and peppermint into individual plastic pots so that they can be plunged into the container separately – the lemon verbena so that it can be lifted and given winter protection, and the peppermint to prevent it swamping the other plants.

3 Plunge the potted lemon verbena and peppermint towards the back of the main pot and infill with compost. Fill the gap at the front of the pot with the camomile. If the peppermint gets too large, lift it out, cut it back, split it and repot it. It will quickly regrow, and the freshest, youngest leaves are best for herbal teas.

Camomile tea

To make 1 cup put 3 camomile flowerheads, a small flowering spray of lemon verbena and a couple of lemon verbena leaves in a cup with 1 teaspoon honey. Add 200 ml (7 fl oz) boiling water and leave to infuse for 4–5 minutes. Don't infuse the herbs for too long or the tea might become bitter.

Pop and go

The sweetcorn cultivar 'Red Strawberry' produces bright red cobs, which are useful for autumn to winter decoration and for edible popcorn. The sweetcorn is underplanted with a carpet of strawberries in an pewter-coloured planter. Make sure the container is in a sunny position, or the sweetcorn will not ripen. To make popcorn the cobs need to be perfectly ripe and thoroughly dried.

Ingredients

1 packet 'Red Strawberry' popcorn seed (or similar red cultivar)

small plastic plant pots

seed compost

drainage material (see page 16)

1 pewter-coloured, square, zinc planter 40 x 40 cm (16 x 16 in)

multi-purpose soil-less compost

4 perpetual-fruiting strawberry plants (such as 'Flamenco')

Method

1   Sow the popcorn seeds in in seed compost in individual pots in spring and place them in a cold frame or greenhouse to germinate and grow on.

2   Once the popcorn plants are large enough to plant out and all danger of frost has passed, add a layer of drainage material to the base of the zinc planter and fill with a good-quality, soil-less compost, select four of the strongest plants and plant them in a square formation.

3   Infill around the popcorn with young, virus-free strawberry plants. Remove all the flowers that form before early summer so that the strawberries produce a more substantial and prolonged crop later in the season.

Strawberry fool

Purée 200 g (7 oz) strawberries with 125 g (4 oz) mascarpone cheese and 1 tablespoon finely grated lemon rind and 1 tablespoon lemon juice. Transfer to a bowl. Whip 150 ml (¼ pint) double cream with 3 tablespoons caster sugar, then fold into the purée. Spoon into 4 bowls and chill until ready to serve. Just before serving decorate with redcurrants, blueberries and extra grated lemon rind.

Gorgeous gourds

Most gourds are edible while they are young and tender, although some are more palatable than others. When left to mature they become ornamental and can be picked and used for indoor decoration. The gourds are grown to climb up an obelisk of copper pipes, intertwined with handsome, yellow-leaved golden hop.

Ingredients

1 packet mixed trailing/climbing gourd or squash seed

small plastic plant pots

seed compost

1 square, bronze-coloured, zinc container, 40 x 40 cm (16 x 16 in)

drainage material (see page 16)

multi-purpose soil-less compost

1 *Humulus lupulus* 'Aureus' (golden hop) plant

1 metal obelisk, 1.2 m (4 ft) high

Method

1 Sow the gourds in spring in seed compost in individual pots and place them in a heated propagating case or greenhouse to germinate. Alternatively, they can be sown outdoors once all danger of frost has passed.

2 Add a layer of drainage material to the base of the zinc container and fill with multi-purpose soil-less compost. Plant out the hop and three of the strongest, healthiest gourd plants, one at each corner of the pot.

3 Insert a pyramidal obelisk framework into the compost and wind the climbing stems of the hop and gourds in as they grow. You can buy a proprietary frame or make your own using copper plumbing pipe. Water copiously, feed once a week and remove yellow or damaged leaves as they appear.

Squash and ginger ice cream

Cut the flesh of a small squash into pieces, steam for 10–15 minutes until tender, then leave to cool. Blend the squash with 4 tablespoons lime juice and 2 pieces crystallized ginger. Put 4 egg yolks and 125 g (4 oz) light muscovado sugar in a bowl over simmering water and whisk until thick, then fold in the squash purée. Whisk 250 ml (8 fl oz) double cream into soft peaks and fold into the squash mixture. Freeze overnight.

Pear delight

tomatoes only

When underplanted with striking, red-leaved radicchio and dwarf, yellow-fruited tomatoes, this columnar pear tree makes a worthy centrepiece for any patio, terrace or courtyard. Although it will crop far better if another pear is planted nearby, 'Concorde' is partly self-fertile and so will produce a respectable crop, even in isolation.

Ingredients

1 packet of tomato 'Yellow Balconi' seed or 3 plants

seed compost

module trays

1 packet of radicchio 'Palla Rossa Red Devil' seed

1 plain terracotta container, 45 x 50 cm (18 x 20 in)

drainage material (see page 16)

1 compact column pear 'Concorde' on Quince A rootstock

soil-based potting compost

multi-purpose soil-less compost

Method

1 Sow the tomatoes in seed compost in module trays in a heated greenhouse or propagating case in early spring or buy young plants in mid- to late spring. Sow the radicchio in seed compost in module trays in a cold frame from mid- to late spring.

2 Site the container in a sunny, sheltered position. Make sure that there are plenty of drainage holes in the base of the container and add a layer of drainage material. Pot up the pear using a half-and-half mix of soil-based and soil-less composts.

3 Once all danger of frost has passed, plant out three strong tomato plants – one on each side of the container, towards the edge – and fill between them with radicchio.

Vanilla pears

Put 1 split vanilla pod in a saucepan with 600 ml (1 pint) vin santo. Stand 6 peeled pears in the saucepan, cover and poach for about 25 minutes. Leave to cool, then transfer the pears to a serving dish. Scrape the seeds from the vanilla pod and add them to the liquid in the saucepan. Boil to reduce to 300 ml (½ pint). Mix with 2 teaspoons arrowroot in a little water and heat, whisking, to thicken. Stir in 1 teaspoon vanilla extract. Leave to cool and pour over the pears.

Passion fashion

In hot summers, the breathtaking blooms of passionflowers are followed by equally eye-catching fruits, many of which are edible. We've used a particularly decorative cultivar with bright red, cherry-like, sweet-tasting fruits. It has climbed over a hooped bamboo cane at the back of the pot, and is accompanied by dwarf Cape gooseberries.

Ingredients

1 packet dwarf *Physalis peruviana* 'Little Lantern' (Cape gooseberry) seed or 3 plants

module trays

seed compost

9 cm (3½ in) plastic plant pots

1 *Passiflora foetida* var. *hirsutissima* or *P. edulis* (passionflower) plant

1 earthenware pot, 40 x 40 cm (16 x 16 in)

drainage material (see page 16)

soil-based potting compost

1 hooped bamboo cane support, 60 cm (24 in) high

Method

1 Sow the Cape gooseberries in seed compost in module trays in a heated greenhouse or propagating case in spring. Once germinated, pot up the strongest seedlings into individual pots.

2 Once all danger of frost has passed, add a layer of drainage material to the base of the container, fill with soil-based compost and place the passionflower at the back and the Cape gooseberries in a group of three in front.

3 Set the bamboo support at the back of the container and weave the passionflower stems around it. Continue to wind the stems in as they grow. After fruiting, the Cape gooseberry plants should be discarded, but the passionflower can be potted up and moved into a frost-free greenhouse or conservatory.

Passionfruit and lime posset

Heat 450 ml (¾ pint) double cream with 50 g (2 oz) caster sugar, then boil without burning. Remove from the heat and beat for 2 minutes. Beat in the juice and grated rind of 2 limes and 2 tablespoons sherry. Chill for 1 hour. Break up the flesh of 4 passionfruits and fold into the cream. Spoon the mixture into four serving dishes and chill for 2 hours. Dust with icing sugar, decorate with passionfruit seeds and serve.

What to grow

If you are determined enough, you can persuade just about any edible plant to grow in a container. Some plants positively thrive in pots, hanging baskets and windowboxes, even performing better than those planted in the open ground. Edible plants are easier to manage when they are cultivated in containers, and there are dwarf or compact cultivars that are specially bred for growing in more confined spaces. There are also some vegetables that can be harvested early, before they outgrow their allotted space.

Vegetables **138**
Fruit **148**
Herbs **151**
Edible flowers **156**

Vegetables

Edible leaves and stems

Cabbage Larger types of cabbage (*Brassica oleracea* Capitata Group) are not suitable for growing in containers. Instead, choose miniature kinds that form small heads and can be harvested and used before they are fully mature. It is possible to have different types of cabbage over a long period: spring cabbages, sown in autumn, can be eaten in spring as loose leaves (spring greens) or when the hearts firm up; summer and autumn cabbages, sown in spring, will mature in four to six months; and winter cabbages, sown in spring, are eaten in winter. Close spacing by sowing at distances of 15–20 cm (6–8 in), rather than the traditional 25 cm (10 in) or more, will give heads about 8 cm (20 in) across. Among the cultivars suitable for this treatment are the early-maturing 'Hispi', the autumn 'Minicole', the Savoy 'Protovoy' and the red cabbage 'Primero'. Cabbages grown in containers need fertile, well-drained, alkaline soil, and they must be watered regularly. Cabbages are susceptible to clubroot and brassica white blister and may be attacked by cabbage root fly, mealy cabbage aphids and flea beetles. Caterpillars are a serious pest of all brassicas so be vigilant and pick them off by hand if spotted.

Cauliflower As with cabbages, standard types of cauliflower (*Brassica oleracea* Botrytis Group) grow too large and take too long to mature to be an economical use of valuable space in a container. Again, however, there are cultivars that can be planted more closely and harvested when they reach the size of a tennis ball or slightly larger. There are also more colourful, ornamental-looking forms with purple, green or even orange-tinted curds. Sow seed in modules in late spring under cover or in early summer outdoors. Take care when transplanting because cauliflower growth is easily checked. Cauliflowers need fertile, moisture-retentive, alkaline soil. They will produce small heads, about 8 cm (3 in) across, when they are set out at distances of about 15 cm (6 in). Sow seed in succession and

they should crop after about ten weeks. Among the cultivars suitable for close spacing in containers are 'Idol' and 'Candid Charm'. Clubroot is the most serious disease, and cabbage root fly, whitefly and caterpillars can cause problems.

Chard or leaf beet Cultivars of chard or leaf beet (*Beta vulgaris* Cicla Group) are among the most colourful and striking of all leafy vegetables. They can be grown as cut-and-come-again crops but should be sown thinly because they do not grow well if overcrowded. Sow seed under cover in early spring or outdoors in mid-spring. They need fertile, well-drained, alkaline soil and must be watered regularly to prevent bolting (running to seed). Swiss chard, also known as silver chard and seakale beet, grows to about 45 cm (18 in) tall and has attractive dark green leaves with fleshy white ribs; cultivars include 'Lucullus', 'Fordhook Giant' and 'White Silver'. Rhubarb chard has dark green, crinkled leaves and dark red stalks and veins; 'Feurio' is more bolt resistant than some. 'Bright Lights' has red, pink, yellow, orange and white ribs. Perpetual spinach or spinach beet has small, dark green leaves borne over a long period; this is a good substitute for spinach although with a slightly coarser texture. Beet leaf miner is the main pest.

Endive Endive (*Cichorium endivia*) are a relative of chicory and can be harvested throughout much of the year. When treated as a cut-and-come-again crop, their leaves are used in salads and have a slightly bitter taste unless blanched by placing an upturned pot over the plants to exclude light. The hardier. broad-leaved kinds are used for winter crops and the more decorative curly-leaved (frisee) endives for summer salads. Sow seed under cover in mid-spring or outdoors in early summer for a late summer crop, in summer for an autumn crop and, if you can protect the plants with cloches, in late summer for a winter crop. Endives do best in fertile, moisture-retentive soil, and they prefer fairly cool conditions. They generally crop 7 –13 weeks after sowing. Curly-leaved cultivars include the very hardy

Cabbage **Cauliflower** **Rhubarb chard** **Endive**

'Green Curled' (or 'Moss Curled'), 'Green Curled Ruffec', 'Kentucky' and 'Sally'. Broad-leaved types include 'Batavian Green', 'Cornet d'Anjou' and 'Eminence'. Slugs and aphids are the most usual problems.

Kale or borecole Kale (*Brassica oleracea* Acephala Group), a relative of cabbage, is grown for the curly edged leaves, which are a useful year-round crop, especially in winter. Red-leaved varieties intensify in colour as temperatures drop. Sow seed in modules under cover in late winter or outdoors in mid- to late spring, transplanting into well-drained soil in full sun. It is possible to harvest leaves of some cultivars seven weeks after sowing. For containers, choose more compact varieties or harvest as a cut-and-come-again crop. Seedlings can be cut when they are 5–8 cm (2–3 in) high and used in salads or the individual leaves picked around 18 weeks from sowing and steamed or stir-fried. Among the best dwarf cultivars are 'Showbor', which has tightly curled leaves; and 'Dwarf Green Curled', which has blue-green, curled leaves. If you have a large container try 'Redbor', the tall 'Black Tuscany' ('Nero di Toscana'), 'Darkibor' or 'Pentland Brig'. Kales suffer from the same pests and diseases as cabbages (see page 138).

Lettuce Of all the salad crops, lettuce (*Lactuca sativa*) is one of the easiest to grow in containers. There is a wide range of leaf shape, colour, texture and taste, and the frilly and red-leaved kinds are particularly decorative. Nonhearting, loose-leaf types are ideal for containers because they can be picked little and often rather than harvested hole. Given the protection of

cloches during the colder months they can be grown virtually year round and are a useful crop for combining with spring bulbs in containers. Sow seed successively from late winter (under cover) to early autumn to give new plants almost all year round. For containers look out for the cultivars 'Blush', the outer leaves of which are tinged with pink, 'Mini Green', a miniature crisphead, which can be set as close as 13–14 cm (4½–5 in) apart, and loose-leaf varieties such as frilly edged 'Fristina and claret-red 'Revolution'. Slugs and aphids are the main pests, and lettuces are also susceptible to downy mildew and grey mould.

Oriental greens Fast-growing oriental greens respond well to the cut-and-come-again treatment both as seedlings and at the semi-mature stage, and their ultimate size is controlled by their spacing – the more space they have, the larger they will grow. These are cool-season crops, which tend to bolt in hot, dry weather, and they are best sown in mid- to late summer for autumn and winter harvest. They need fertile, moisture-retentive soil and full sun. They can either be sown direct into the container, where they will germinate rapidly, or be raised in module trays and transplanted when they are large enough to handle. **Chinese cabbage** (*Brassica rapa* var. *pekinensis*) matures 8–10 weeks from sowing and is easy to grow in containers. It forms heads of bright green leaves, varying in shape from barrel-like to cylindrical. There are loose-leaf types that are less prone to bolting. **Mizuna greens** (*B. rapa* var. *nipposinica*) are a good, low-temperature crop. The deeply cut, dark green leaves have a fresh, crisp taste and can be used raw in salads, cooked with meat dishes or pickled. Mizuna grows well between sweetcorn as a cut-and-come-again crop. **Pak choi** (*B. rapa* var. *chinensis*) is an easy-to-grow vegetable for containers with bright green leaves that have prominent white veins. The height varies depending on cultivar, and they can either be harvested young or thinned out and left to mature. The small, neat Shanghai types are best suited to containers. **Spinach mustard** or komatsuna (*B. rapa* var. *perviridis*) is an exceptionally fast-growing plant, bearing large, bright green, sometimes red-tinged, leaves with a mild but distinct mustard flavour. The main pests of all these plants are flea beetles and slugs.

Radicchio A form of chicory, radicchio (*Cichorium intybus*) forms an attractive plant with glossy green leaves that may be tinged with bronze and red, the colour spreading and deepening as they mature to form crisp, succulent round heads. Radicchio is very cold tolerant so is the perfect winter crop. Sow seed outside in early summer for late-summer picking or in late summer for a winter crop. 'Palla Rossa' produces tight heads of leaves that turn redder as the temperature drops; it can be grown for its leaves, hearts or forced for chicons. 'Rossa di Treviso' has red-and-green variegated leaves. 'Rossa di Verona' has deep red leaves that form tight heads; if you cut plants back in autumn and protect them with a cloche they sometimes produce a second head in spring. Slugs may be a problem, but radicchio is usually trouble free.

Rocket Also known as argula, rucola or roquette, rocket (*Eruca versicaria*) is an excellent cut-and-come-again crop, providing a supply of spicy, fresh leaves over a long period from spring to early winter. It is a fast-growing, half-hardy annual, 60–90 cm (2–3 ft) tall. Both the leaves and pale yellow flowers can be used in soups and salads and as a garnish. Sow seed in succession from spring onwards in moisture-retentive, rich soil in partial shade. Leaves can be cut for the table as soon as three weeks after planting. Flea beetles may be a problem, and in hot weather plants often bolt.

Salad burnet This easily grown, evergreen perennial herb (*Sanguisorba minor*) is completely hardy. Plants, which grow to 1 m (3 ft) tall, have attractive, bright green leaves, divided into oval leaflets, and spikes of red flowers in summer. Sow seed in spring and grow in any well-drained soil in sun or partial shade. The leaves, which have a rather nutty flavour with a hint of cucumber, can be harvested at almost any time of year and are particularly useful for winter salads. Slugs can be a problem, but salad burnet is otherwise trouble free.

Salad leaves In addition to well-known salad leaves, there is a huge and ever-expanding range of plants with interesting, often colourful, edible leaves that can be added to salads to give a variety of flavours. You can buy packets of seeds containing mixtures of edible leaves. Look out for mixtures such as 'Saladini' and 'Mesclun' (misticanza) in seed catalogues. Sow

Lettuce

Red amaranth

Radicchio

Variegated land cress

seed outdoors in early to mid-spring for summer cropping or in mid- to late summer for winter cropping under cover. These plants need fertile, well-drained but moisture-retentive soil. Water them regularly to prevent bolting. The group includes **Amaranths** (*Amaranthus* spp.), which are more often grown as ornamental plants, but some species and varieties have tasty leaves that are a worthy substitute for spinach. Some also produce nutritious edible seeds. Red amaranths have ornamental leaves tinted red or purple. **Corn salad**, also known as lamb's lettuce or mache (*Valerianella locusta*), has a low, carpeting habit, and is an ideal cool-weather salad green for relatively shallow containers and will provide a virtual year-round crop of small green leaves. Although it does not bolt easily, its flavour diminishes in warm weather. Corn salad is at its most flavoursome when young, so make small successional sowings every couple of weeks from early spring. **Mustard** (*Brassica hirta*) germinates and develops quickly and is an ideal cut-and-come-again crop. Sow seed thickly in spring or early summer and keep cool and well watered for a super-fast crop. Red-leaved cultivars are particularly attractive. There are green-, red- and golden-leaved cultivars of **orache** or mountain spinach (*Atriplex hortensis*), which tastes much like spinach. The colourful young leaves can be added to salads, but more mature leaves are best steamed before eating. When left to grow, orache can reach 1.5 m (5 ft) or more in height, so nip out the growing tips to encourage branching and use them in salads. Sow orache in late spring or early summer when all danger of frost has passed. **Para cress** or Brazil cress (*Acmella oleracea*) is an unusual but attractive trailing annual, bearing bronze-green new shoots and leaves, coupled with small, round, yellow-bronze flowers. It is easy to grow from seed. Add the young leaves and shoot tips to mixed salads for extra flavour. **Summer purslane** (*Portulaca oleracea*) is a fleshy-leaved salad plant, and the golden-leaved form, *P. oleracea* var. *aurea*, is particularly ornamental. The leaves can be picked 4–8 weeks after sowing. They have a sharp taste and are best used sparingly in salads or cooked. The stems and leaves can be pickled and used in winter.

Sorrel A hardy perennial, sorrel (*Rumex acetosa*) forms a rosette of basal leaves above which long stems with long-stalked leaves and spikes of small red flowers are borne, reaching 60 cm (24 in) or more tall. There are several variants, of which blood-veined sorrel is the most decorative, with its deep green leaves prominently veined dark red. Sow seed in spring or autumn, spacing plants 12 cm (30 in) apart, and remove the flower spikes to encourage leaf production. Grow in deep, slightly acid, well-drained soil in sun or partial shade. French or buckler-leaved sorrel (*R. scutatus*) is a neat, low-growing leafy green with a less pungent flavour, which can be added to soups, salads and used as a seasoning for seafood, potato dishes and rice. Apart from slugs, sorrel is trouble free.

Spinach Spinach (*Spinacia oleracea*) is a highly nutritious, cool-weather crop, and it is often said to taste better after a frost. It is fast and easy to grow in containers and is particularly useful for sowing among slower-maturing crops because the spinach will be harvested long before the other crop matures. Sow seed outdoors in early to late spring, and in late summer and early autumn for a succession of plants. They need fertile, well-drained, alkaline soil. For the best flavour keep the leaves tender and hydrated by ensuring the compost is evenly moist. In hot weather spinach has a tendency to bolt, so plants should be harvested as soon as they show signs of doing so to save the crop. 'Teton' is a compact variety and 'Bordeaux' has attractive red stems. The main problems are slugs and downy mildew.

Winter cress or land cress A good winter salad crop, winter cress (*Barbarea vulgaris*) has peppery-flavoured leaves, which both look and taste somewhat like watercress. The variegated form *B. vulgaris* 'Variegata' has leaves splashed creamy-yellow. Leaves can be harvested all year round, but plants are best covered with cloches to keep them in good condition through the winter. Plant in fertile, well-drained soil in full sun or partial shade. Seed sown in mid- to late summer can be cropped from winter to spring. Water regularly to prevent plants from bolting. Flea beetles are the main pests.

Edible roots and tubers

Beetroot or beet Beetroot (*Beta vulgaris* subsp. *vulgaris*) is mostly grown for its edible root, which is commonly globe-shaped, but sometimes cylindrical or tapered, but the leaves, especially when young, are also edible and make a tasty addition to salads. Sow seed under cover from late winter to early spring or outdoors from early spring to mid-summer for a succession of plants. Grow beetroot in well-manured, well-drained soil and in an open, sunny position. Water regularly to prevent the roots from turning woody and bitter. Most beetroot 'seed' actually produces several seedlings, which makes thinning and spacing difficult. However, for a good crop from a small space, beetroot can be multi-seeded, which means growing 6–8 seedlings in a single pot and planting them together. As they grow they produce a clump of small but perfectly shaped roots. If you prefer to grow individual plants look out for monogerm seed. There are numerous cultivars. 'Kestrel F1', 'Monaco' and 'Detroit 2 – Little Ball' are suitable for growing as mini-beets, but 'Bull's Blood', which has dark red leaves, 'Chiogga', with its candy-striped roots, and 'Boltardy', with its dark red roots, are reliable and widely available. Apart from bolting in hot weather, beetroot are trouble free.

Carrot Unless you are growing them in particularly deep containers, such as the drainage pipes in the Towering Thymes recipe (see pages 40–41), it is best to choose round or stump-rooted varieties of carrot (*Daucus carota* var. *sativa*) for a container because they do not require such a depth of compost. Carrots do not transplant well, so sow seed under cover in biodegradable pots or module trays in late winter or outdoors in early spring. Carrot seeds are tiny, so are best sown in small pinches and thinned out to about 2.5 cm (1 in) apart in order to allow the roots to develop. To avoid root disturbance, thin out by snipping off the unwanted seedlings at soil level rather than pulling them out. Make successional sowings every few weeks from early spring to late summer for a continuous supply of tender young roots. Carrots dislike heavy, rich soils, preferring free-draining, sandy compost. 'Early French Frame' and 'Parmex' produce round roots, which are best eaten when they are the size of golf balls. 'Redcar' and 'Chantenay Red Cored' develop short, stumpy roots, which are also ideal for containers. The most common pest is carrot fly. However, these fly close to the ground and are less likely to attack carrots raised above soil level in containers.

Potato Early potatoes (*Solanum tuberosum*) are best suited to containers because they crop quickly and have less top-growth than maincrop types. Plant the tubers in late spring in as large and as deep a container as you can find and have space for, setting them about 10 cm (4 in) apart on a 10–12 cm (4–5 in) bed of multi-purpose compost. Cover them with a 5–8 cm (2–3 in) layer of compost. Shoots will emerge within a couple of weeks, and when they are about 15 cm (6 in) high, add more compost, half-burying them. Repeat this process a few times through the growing season as potato tubers grow from the part of the stem that is below soil level, and this will encourage them to crop well. Keep them well watered but not overwet, which encourages disease. When grown as 'new' or salad potatoes, they can usually be harvested around two months or so from planting. Otherwise leave them to flower and allow the top-growth to wither before tipping them out of the container. Choose cultivars such as 'Accent', 'Mimi' and 'Swift'. The main problem of potatoes is blight, which can devastate crops, but this affects plants in summer in periods of warm, wet weather and is unlikely to affect earlies grown in containers.

Radish Radishes (*Raphanus sativus*) are fast maturing and ideal for growing among other crops that take longer to develop. They will grow in sun or light shade, but when they are grown in warmer conditions the roots tend to have a stronger, more peppery taste. The roots are generally rounded and form just below soil level, so these are ideal crops for shallow containers. They can also be spaced fairly closely so it is possible to grow a good crop in a relatively small container. Radishes are best sown direct every couple of weeks from early spring to early

Beetroot

Potato

Radish

Carrot

autumn for a succession, but they can also be raised in module trays and transplanted when they are large enough. They do best in well-drained, fairly rich, alkaline soil. There are many cultivars to choose among, with roots of varying colours and shapes. 'Helro', 'Cherry Belle' and 'Sparkler' have round, red-skinned roots, and 'Pink Beauty' has round, pink-red roots. 'Black Spanish Round' is, as the name suggests, a round, black-skinned, white-fleshed radish. Flea beetle and cabbage root fly can be problems, and radishes are also susceptible to clubroot.

Turnip Turnips (*Brassica rapa* Rapifera Group) are most familiar as the large root vegetables, but there are numerous smaller cultivars, which can be harvested when they are small and tender and used immediately. These turnips do not store well but are ideal as a side vegetable, steamed or sautéed, for use in summer soups or grated in salads. The leaves can also be cooked or, when young, eaten raw. The roots can be harvested when they are 2.5 cm (1 in) across, so they can be sown fairly close together and gradually thinned out. Like radishes, they are best sown directly into the container, around other crops such as sweetcorn or peas, but they can also be raised in module trays and transplanted. They need fairly rich, well-drained soil and regular watering. 'Arcoat', a summer turnip that can be grown as a catch crop, has red-topped, white roots. 'Tokyo Cross', also a summer crop, is a fast-growing cultivar, best sown from early summer to prevent bolting. Turnips are prone to many of the same pests and diseases as cabbages (see page 138). They are also susceptible to violet root rot and suffer from turnip gall weevils.

Edible bulbs and stems

Garlic Garlic (*Allium sativum*) grows surprisingly well in containers, provided they are positioned in a warm spot and are watered regularly. Garlic needs a long growing period for best results, and to maximize space while they are maturing intersperse them with fast-growing salad leaf crops (see page 140). Plant individual cloves (pointed end up) in late autumn or, in mild areas, in late winter to early spring. They need free-draining, fertile, alkaline soil that must never dry out. The range of cultivars has increased in recent years. 'Cristo' produces large bulbs, whether it is sown in autumn or spring. Also suitable for autumn or spring planting are the strongly flavoured 'Fleur de Lys' and 'White Pearl'. 'Long Keeper' does best from autumn planting, but 'Printanor' and the mild 'Sultop' are good choices for spring planting. Onion white fly and leek rust are the main problems.

Kohl rabi The edible part of kohl rabi (*Brassica oleracea* Gonglyodes Group) is the swollen stem that develops just above soil level and tastes like cauliflower. They can be used steamed or boiled with other vegetables or grated raw in salads. Kohl rabi is best harvested when the roots are the size of billiard balls, 8–10 weeks from sowing, before they become tough and woody. Sow seed under cover from late winter or outdoors from early spring, sowing at two-week intervals to give a succession of plants. Grow in well-drained, light soil in full sun, setting seedlings 20 cm (8 in) apart. Among the best cultivars for growing in containers are 'Logo' and

'Rolando'. F1 hybrid 'Kolibri' is purple-skinned and 'Green Delicacy' is a pretty pale green. Although they are susceptible to the same pests and diseases as other brassicas, it is usually harvested before it can succumb to them because kohl rabi is so quick growing.

Leek Leeks (*Allium ampeloprasum*) are generally thought of as staple plants in the vegetable plot or allotment, but they are one of the vegetables that can be grown successfully as mini- or baby vegetables. Sow seed direct from early spring to early summer, thinning seedlings to 1 cm (½ in) apart as they emerge. The plants will grow quickly as long as they are in fertile, well-drained soil and are watered regularly. Pick them when they are as thick as a pencil, pulling them in bunches. Grown in this way, leeks are an excellent substitute for spring onions and can be used in salads or braised. If you allow them to grow on, they can be pulled when they are about 1 cm (½ in) in diameter. 'King Richard' and 'Jolant' are the best cultivars to grow in this way. Leeks suffer from onion fly maggots and are susceptible to white rot and rust.

Onion Traditionally grown onions (*Allium cepa*) are too large for most containers, but most cultivars can be harvested before they are fully mature. It is also possible to multi-sow onions, putting 6–8 seeds together in individual pots or modules. Instead of thinning the plantlets, transplant the whole pot, setting the clumps about 15 cm (6 in) apart. They will develop into small, perfectly round onions. Seed can be sown under cover from late winter or early spring and outside from mid-spring. Sets (immature onions) can be planted where you want them to grow but generally become too large for most containers, although the closer you plant them the smaller they will grow. Onions need well-drained, alkaline soil in an open, sunny position. There are many cultivars, of which 'Shakespeare', 'Imai' and 'Shimonita' are particularly suitable for growing as mini-onions. The main pests are onion fly and onion thrips. Onions are also susceptible to several diseases, including onion neck rot and white rot.

Shallot Shallots (*Allium cepa* Aggregatum Group) have a milder, more delicate flavour than onions. When they are grown from sets they form multiple bulbs, but seed-grown plants produce single bulbs. Seed can be sown outdoors from early spring, while sets are usually planted from late winter. Seeds can be multi-sown, with 6–8 seeds to each pot and planted out in clumps. Shallots need the same conditions as onions (see above). Immature shallots can be harvested and used in the same ways as spring onions. 'Ambition' and 'Matador' are usually grown from seed; 'Delicato', 'Pikant', 'longer' and 'Red Sun' are grown from sets. Shallots suffer from the same pests and diseases as onions.

Podded vegetables

Asparagus pea An unusual half-hardy annual is the asparagus pea (*Lotus tetragonolobus*), which bears small, winged (four-sided), edible pods containing smooth, brown seeds. The whole pods are usually picked when they are about 2.5 cm (1 in) long and steamed and served with butter. The common name derives from the fact that they taste of asparagus. Sow seed under cover in early spring or outdoors in late spring, after the last frost. The seeds develop into small bushes with trailing stems, to 40 cm (16 in) long. Bright red flowers are borne in summer, and the pods follow the flowers, 8–10 weeks after sowing. These plants are trouble free.

French bean One of the mainstays of the vegetable plot are French beans (*Phaseolus vulgaris*), which are increasingly included in borders and ornamental plantings because of their attractively coloured flowers and fruit. Get plants off to an early start by sowing seed in biodegradable pots under cover in mid-spring; harden off and plant out only after the last frost. Alternatively, sow direct in early summer. Plant into fertile, well-drained, alkaline soil and grow in full sun. There are several dwarf cultivars, which are particularly suitable for containers, including 'Aramis', 'Arosa', 'Cropper Teepee', 'Ferrari', 'Maxi', 'Masai', 'Mont d'Or', 'Safari' and 'Sungold'. 'Purple Teepee'

Kohl rabi

Shallot

Asparagus pea

Runner bean

and 'Golden Teepee' bear attractive purple and yellow pods, respectively. The main pests are bean seed fly, black bean aphids and slugs.

Hyacinth or dolichos beans Hyacinth beans (*Lablab purpureus*) are grown for the young pods and seeds (pulses), which can be used both fresh and dried. In temperate areas this tender perennial is grown as an annual, and it needs some support because it is a naturally climbing plant. Sow seed under cover in spring and do not plant out until the temperature is about 18°C (64°F). The white or purple flowers are followed by green or purple pods, to 15 cm (6 in) long, which contain up to six seeds, which may be white, cream, reddish, brown or black. They are so decorative that they are often included in the flower section of seed catalogues. Although tender, these plants are trouble free.

Pea There are few things that equal the flavour of freshly shelled peas (*Pisum sativum*), quickly cooked and served with butter and a little mint. As long as you have the space and choose cultivars carefully it is possible to have peas from late spring to early autumn. However, if you are using a single cultivar for a container, make several sowings of seeds at two-week intervals, sowing under cover from early spring and outdoors as soon as the soil is warm and dry (peas will not germinate in cool, wet conditions). Grow them in as deep a container as possible, using well-drained, alkaline soil. Water regularly but avoid overwetting the compost. 'Feltham First' and 'Meteor' are compact varieties and 'Half Pint' is very dwarf. 'Sugar Dwarf

Sweet Green' ('Norli') is a mangetout, and 'Sugar Rae'. 'Sugar Ben' and 'Sugar Gem' are compact sugar snap peas. Start harvesting pods from the bottom of the stem. Peas are susceptible to powdery mildew and are sometimes infested by pea moths, which burrow into the peas inside the pods so the damage is usually unnoticed until you start to shell the peas before cooking them.

Runner bean Although they are really perennials, runner beans (*Phaseolus coccineus*) are grown as annuals in most temperate gardens. They are mostly familiar in the garden when they are trained up wigwams of bamboo poles, but dwarfer cultivars are available and need no support. Sow seed under cover in late spring into biodegradable pots and harden them off for planting out after the last frost. Runner beans need moisture-retentive, alkaline soil. To turn climbing runner beans into bushy, more manageable plants, pinch out the growing tips when they are around 30 cm (12 in) high and remove any climbing shoots as they appear. The cultivars 'Kelvedon Marvel' and 'Scarlet Emperor', normally climbing forms, can be grown as bushes in this way. Among the true dwarf cultivars, which do not need staking, are the stringless 'Pickwick' and 'Hammond's Dwarf Scarlet', a widely grown, red-flowered cultivar, which will crop over a long period if the beans are picked regularly. The dwarf 'Hestia' has red and white flowers and is recommended for containers. Slugs and aphids are the main problems with runner beans, but in cold or dry conditions or when there are few flying pollinators failure to set pods can lead to a reduced crop.

Aubergine   Squash   Tomatoes   Chilli pepper

Fruiting vegetables

Aubergine (eggplant) Among the most attractive of container-grown vegetables are aubergines (*Solanum melongena*), which bear fruit in a surprising range of shapes, colours and sizes. Aubergines do not grow well if there are too many other plants in the pot around them, but with a little coaxing and extra care they can usually be persuaded to give up space to low-growing crops, such as dwarf beans. Sow seed in early spring in a heated propagating case. Given plenty of heat they will germinate well, but they resent root disturbance so are best sown in small individual pots. Grow them on in a warm, protected environment before hardening off and planting out after the last frost. Aubergines need fertile, well-drained but moisture-retentive soil. Water regularly and apply a high-potash fertilizer as soon as fruits appear. Pick crops as they ripen to encourage the production of new fruits. 'Fairytale F1' has attractive striped fruits, and 'Baby Rosanna F1' bears an abundance of dark purple fruits and has purple-tinged foliage. The fruits of 'Snowy' are long and white, and 'Bonica' is a compact form with large, purple fruits. Irregular and inadequate watering can cause blossom end rot, and the worst pests are aphids and, in dry weather, red spider mites.

Courgette (zucchini) Traditionally, these half-hardy annuals grew too large for containers, but new cultivars of courgette (*Cucurbita pepo*) have made it possible to get a worthwhile crop from pot-grown plants. Sow seed under cover in late spring, sowing two seeds to each pot and removing the weaker seedling if both germinate. Plant out after the last frosts into fertile, moisture-retentive soil. These are greedy plants and need regular watering and feeding. Choose cultivars such as round, yellow-skinned 'One Ball F1' which is best harvested at tennis ball size, or 'Supremo', 'Patriot' and 'Bambino' which should be harvested when the fruits reach 15 cm (6 in) in length. Powdery mildew affects plants in hot, dry weather, and courgettes are also susceptible to mosaic virus, which is spread by aphids. Slugs are the main pest.

Cucumber and gherkin To crop well, cucumbers and gherkins (*Cucumis sativus*) need a plentiful and consistent supply of food, water and heat. They are climbing or trailing plants, and even those described as compact or semi-bush will still scramble around to some extent. They will require some form of support, such as trellis or a wicker pyramid, or they can be planted in a tall container and allowed to cascade over the sides. Sow seed into individual biodegradable pots, two seeds in each, in a heated propagating case in spring. Remove the weaker seedling if both germinate. To encourage sideshoots, pinch out the main growing tip once seven or eight leaves have formed. Plant out into fertile, moisture-retentive soil after the last frost. They need a warm, sheltered spot in full sun. Although it is tempting to let cucumbers grow as large as possible, the smaller fruits are usually more succulent and have the best flavour. Pick regularly to encourage the formation of more fruits. Outdoor cultivars include 'Burpless Tasty Green', 'Marketmore', which is resistant to virus diseases, and 'Bush Champion', which is a compact plant with dark green fruits. Older cultivars include 'Crystal Lemon' and 'Crystal Apple', which produce small, oval, pale yellow fruits that taste just like conventional cucumbers. Gherkins are not small cucumbers, although they are grown in the same way. Look out for cultivars 'Conda' and 'Venio'. Slugs and aphids are the main pests, and plants may suffer from powdery mildew in hot, dry conditions. However the main problem is cucumber mosaic virus, which is transmitted by aphids; always try to buy certified virus-free plants.

Peppers and chillies Tender, annual, sweet or bell peppers (*Capsicum annuum* Grossum Group) and chilli peppers (*Capsicum annuum* Longum Group) can be grown as neat, rather bushy plants that are ideal for containers in a sheltered, sunny part of the garden. Sow seed in a heated propagator in early spring and plant outside only after the last frost into fertile, well-drained but moisture-retentive soil. Like tomatoes, peppers and chillies need feeding with a high-potash fertilizer and regular watering. There are dozens of sweet pepper cultivars, including 'Redskin', a compact form, 'Gypsy', which has pointed fruits that ripen from yellow-green through orange to red, and 'Marvras', which has dark purple fruits, ripening to red. Pick them when they are smooth and ripe and use within about ten days. Chilli peppers range from the very hot 'Habañero' to the milder 'Hungarian Wax' (although this gets hotter as it ripens). 'Apache' is a particularly dwarf, red-fruited cultivar. Surplus chillies can be dried or frozen. The main problem is blossom end rot, the result of inadequate or irregular watering.

Pumpkin, squash and gourd Pumpkins (*Cucurbita maxima*), squashes (*C. moschata*) and gourds (*C. pepo*) are easy to grow plants that have a reputation for being too large for containers. However, there is now such a range of shapes and sizes that it is possible to find cultivars that can be grown in a large pot. Sow seed under cover in early spring and harden the plantlets off before planting out after the last frosts. They need fertile, well-drained but moisture-retentive soil and a position in full sun. Summer squashes will bear fruit seven or eight weeks after planting out, while you can expect to harvest winter squashes and pumpkins 12–20 weeks after planting out. 'Jack Be Little'

produces small orange pumpkins, about 10 cm (4 in) across, which can be roasted. Among the smaller squashes are 'Sunburst' and 'Peter Pan'. The main problems are slugs and, in hot, dry weather, powdery mildew.

Sweetcorn Sweetcorn (*Zea mays*) is not an obvious candidate for containers. However, a warm, sheltered, sunny spot and careful attention to watering and feeding can produce a worthwhile crop in a very small space. Sweetcorn does not do well in cold, wet seasons. Sow seed under cover in biodegradable pots in mid-spring and plant out into warm, fertile, moisture-retentive soil after the last frosts. Choose mini-corn, such as 'Minisweet' or 'Minipop F1', which bear the small baby corns often included in stir-fries. Harvesting while young will help promote the production of new ears. It is possible to cram a lot of corn plants into a large container, but it is usually better to give the plants more space and include another, earlier maturing crop, such as dwarf beans or peas, at their feet. As well as the traditional yellow or white sweetcorn cultivars, there are ornamental corns with coloured cobs, most of which are better for drying and using for decoration rather than being eaten. 'Red Strawberry' is suitable for making popcorn. The main problems are mice, birds and slugs.

Tomato Tomatoes (*Lycopersicon esculentum*) are perfectly suited to containers, and there are now cultivars that are suitable for hanging baskets. Although they can be grown from seed sown under cover in late winter, it is easier to buy plants and set them out after the last frosts. They need a position in full sun and fertile, well-drained soil. Water and apply a high-potash feed regularly throughout the growing season, although overwatering will impair the flavour. Cherry tomatoes, such as 'Cherry Belle', 'Gardener's Delight', 'Nectar' and 'Yellow Balconi', are ideal for containers, while 'Tumbling Tom Yellow' and 'Tumbling Tom Red' are a trailing form, perfect for a hanging basket. The main pests are aphids and whitefly. Inadequate watering causes blossom end rot (a sunken dark patch on the fruit), and outdoor tomatoes are susceptible to potato blight (see page 142).

Fruit

Fruit trees, such as apples and pears, grown on dwarfing rootstocks are ideal for planting in containers and will give a crop sufficient to warrant the space they will take up. As a bonus, they will supply an attractive, if fleeting, display of spring flowers.

Cane fruits are not well suited to growing in containers, but bush fruits can look very attractive in pots and give a worthwhile crop.

Apple When you are choosing an apple tree (*Malus domestica*) to grow in a container make sure it is one that is grown on a dwarfing rootstock such as M27, which will give a plant about 1.8 m (6 ft) high, or MM106, which will give a plant about 3.6 m (12 ft) high. Use the largest container you can afford and have room for – a half-barrel, for example – and fill it with fertile, moisture-retentive but well-drained soil. Only a few apples are self-fertile, so check the pollination requirements of the cultivar before you buy. If you have space for only one apple, look out for 'family' trees, which have three cultivars grafted on to the same rootstock. Also check whether you have chosen a spur-bearing or tip-bearing apple, which will affect the pruning. Make sure the container is in a sunny spot and water regularly so that the compost never dries out. Although you could grow some ornamental annuals in the same container as the tree during the first year, the tree will grow better in subsequent years if nothing competes with it for nutrients and moisture. Apple trees are susceptible to mildew, canker, rust and scab and may be infested by apple sawfly maggots and codling moths, which make holes in the fruits, and by aphids and capsid bugs.

Banana Banana plants (*Musa* spp.) make striking ornamental plants for a sheltered patio or sunny corner, but they are not hardy and need to be moved to the protection of a heated greenhouse or conservatory in winter. The large leaves are easily damaged by strong winds, and even in summer they need a sheltered position. Buy plants from a nursery or divide an established plant in spring. Bananas need fertile, well-drained soil and a position in full sun. Fruits are most likely to be borne on *M. acuminata* 'Dwarf Cavendish', as long as winter temperatures do not fall below 15°C (59°F); it will eventually grow to 3 m (10 ft) tall and across. When they are grown under glass, bananas are susceptible to mealybugs and aphids.

Blueberry Blueberries (*Vaccinium corymbosum*) are one of the best fruits to grow in containers because they need moist, acidic soil, with a pH of 4.0–5.0, which is rarely found in gardens. They are deciduous bushes, which are perfectly hardy, although late frosts can kill the spring flowers. Always use rainwater for watering. Although blueberries are self-fertile, you get a better crop if you have space for two plants. Early cultivars

Apple

Blueberry

Cranberry

Mandarin orange

include 'Bluecrop', which bears large, well-flavoured fruits, which ripen by late summer, and 'Patriot', which has especially large fruits. Mid-season blueberries include the vigorous 'Berkeley', 'Herbert', 'Ivanhoe' and 'Goldtraube'. Late-season cultivars include 'Jersey' and 'Coville'. Prune to encourage the production of new branches because the fruits are borne on two- and three-year-old wood. Remove dead wood in winter as well as branches that are four years old. Blueberries rarely suffer from pests and diseases, although it may be necessary to net plants to protect the fruits from birds. Yellow leaves are a sign of iron deficiency or a too high soil pH.

Cape gooseberry The Cape gooseberry (*Physalis peruviana*) is related to the ornamental Chinese lantern or bladder cherry (*P. alkekengi*), and its small yellow fruits are enclosed in similar papery calyxes. The tender perennials, which grow to 1 m (about 3 ft) tall, can be grown from spring-sown seed, but it is easier to plant roots. Grow in well-drained, fertile soil in full sun. There are several cultivars, including 'Goldenberry' and dwarf 'Little Lantern', both of which have large, golden-yellow fruits. The fruits can be eaten fresh but are more often used for jams and jellies. The plants are trouble free.

Citrus Oranges (*Citrus aurantium*), lemons (*C. limon*) and other citrus fruit can be grown successfully in containers as long as they can be moved to a frost-free position in winter. They are spiny, evergreen shrubs, with glossy green leaves and fragrant, white flowers in spring. Plant in well-drained, fertile soil The best-known lemon cultivar is 'Meyer', which is a compact form with round fruits. Citrus fruits that are grown under glass are often infested with red spider mite, whitefly, scale insects and mealy bug; plants grown outside sometimes suffer from root rot.

Cranberry Like blueberries, cranberries (*Vaccinium oxycoccos*) need moist, acidic soil, and they should be watered with rainwater. They are small, evergreen bushes, from 15 cm (6 in) to 60 cm (24 in) tall, with thin stems and small leaves. The round, red fruits, which should be picked before the first frosts, are used for making cranberry jelly or sauce, the traditional accompaniment for turkey and game. The cultivar 'Olson's Honkers' bears large fruits, and 'Pilgrim' is a fast-growing form. Cranberries are rarely troubled by pests and diseases, but they do suffer from chlorosis if the soil pH is too high.

Grape A grape vine (*Vitis vinifera*) is not the easiest of fruits to grow in a container, but with perseverance you can encourage your plant to produce a worthwhile crop, and a well-grown, well-pruned specimen grown on a short woody stem or 'leg' as a standard will make a handsome addition to a sunny patio. Position the container near a warm wall to shelter the plant. The vine will crop best when it is grown in warm, dry, fairly poor soil, and when it has as much light, space and air as possible. The fruits are borne on new growth, so after fruiting prune shoots back to encourage the formation of new ones for next year. Top-dress each spring with a seaweed fertilizer and give the vine a light liquid feed every two or three weeks during the growing season. Strict, careful pruning will help keep it compact and productive. Moulds and mildews can be a problem, particularly in humid summers.

Passionflower The unmistakable, exotic-looking flowers of *Passiflora* spp. are followed by fruits that are usually edible if, as in the case of *P. caerulea* for instance, not always palatable. The species *P. edulis*, which is sometimes known as purple granadilla, produces fruits that can be used to make juice or to flavour ice cream. This is a tender plant, which needs a winter temperature no lower than 16°C (61°F), so make sure that the container is not so large that you cannot move it to sheltered accommodation in winter. These tendril climbing plants will need a trellis or other support. Plant them in fertile, moisture-retentive but well-drained soil in full sun.

Pear   Passionfruit   Redcurrant   Strawberry

Hardier species, such as *P. caerulea*, will survive outdoors in winter if they are in a sheltered position and protected from cold, drying winds.

Pear If you grow a pear (*Pyrus communis*) in a container make sure that it is growing on a dwarfing rootstock such as Quince A, which will give plants to 3.6 m (12 ft) high, or Quince C, which produces plants about 2.4 m (8 ft) high. Only a few pear varieties are self-fertile or partially self-fertile, otherwise you will need to grow two compatible cultivars. Choose the largest possible container and position it in a sunny position, sheltered from strong, blustery winds. Use fertile, moisture-retentive but well-drained soil. Pears need pruning to remove dead and damaged wood and to encourage a strong central leader to develop. They do not store as well as apples, and some cultivars are best eaten from the tree.

Red-, white- and blackcurrant Red- and whitecurrants (*Ribes* spp.) are hardy deciduous shrubs, which are tolerant of a range of conditions. Grow in any well-drained soil. They are self-fertile plants and form bushes about 1.2 m (4 ft) high. The fruits are borne on shoots that develop from the permanent framework of branches and will start cropping as soon as any branches are two years old. Redcurrant cultivars include 'Stanza' and 'Red Lake', and a reliable and popular whitecurrant is 'White Versailles'. Blackcurrants, which prefer more fertile soil, bear fruit on wood grown during the previous year and so are pruned differently. Among the best cultivars are 'Ben Lomond' and 'Ben Sarek'. All the currants are susceptible to mildew and may need netting to protect the fruit from birds.

Strawberry If you choose cultivars with care, it is possible to have strawberries (*Fragaria* x *ananassa*) from late spring to early autumn. Plant bare-rooted runners in summer or pot-grown plants in autumn, using fertile, well-drained soil in full sun. Because they are shallow-rooting plants, it can be difficult to plant them with other species in the same container, and they are usually mulched, traditionally with straw, to keep down weeds and to keep the fruits above the soil. There are many cultivars to choose among. 'Emily' and 'Rosie' are early season plants, 'Cambridge Favourite' and 'Pegasus' are mid-season cultivars, and 'Florence' and 'Rhapsody' are good, late-season plants. Among the perpetual-fruiting (remontant) plants are 'Bolero', 'Challenger' and 'Elan'. Mould can be a problem, and slugs and snails find strawberries irresistible, as do birds.

Herbs

Many of the most popular and useful culinary herbs thrive in containers, and some, including rosemary and sage, are evergreen, providing a year-round display of aromatic leaves for cooking. When they are in pots herbs can be positioned where they are readily accessible – near the kitchen door, close to a path or on a windowsill – and they can be moved into a more sheltered positions for the winter months, even under glass, giving a supply of fresh young leaves for culinary use throughout the year.

Chive and garlic chive The hollow, onion-flavoured leaves of chives (*Allium schoenoprasum*) are frequently added to salads and egg dishes, and both the leaves and bulbs can be used in soups and sauces. Add the pretty flowers to salads for the distinctive taste of onion. Garlic or Chinese chives (*A. tuberosum*) is similar but has a stronger flavour and flat leaves. Sow seed of both these perennial plants outdoors in spring or divide established clumps in spring. They do best in well-drained, fertile soil in full sun. Water regularly during dry weather. Chives, which grow 30–45 cm (12–18 in) high, bear pink, purplish or sometimes white flowers in summer. Garlic chives, 25–50 cm (10–20 in) high, have white flowers in late summer to autumn. Both are susceptible to the usual diseases of onions: white rot, downy mildew and onion fly.

Lemon verbena Lemon verbena (*Aloysia triphylla*) is a potentially quite large, deciduous shrub, growing to 3 m (10 ft) high and across in left unpruned. In late summer it bears little, pale lilac flowers in panicles. This herb is not reliably hardy and needs the protection of a sheltered patio or a deep mulch over the roots in winter. Lemon verbena is usually propagated from cuttings, so look out for plants in garden centres. Prune the plants in spring, cutting back the main stem to 30 cm (12 in). Lemon verbena does best in full sun in any well-drained soil. Pick the aromatic leaves to use fresh in tisanes or in salads or you can dry them for a delicious, citrus-flavoured tea. As long as they do not get too cold in winter, lemon verbena plants are trouble free.

Dill A hardy annual, dill (*Anethum graveolens*) is grown for its feathery, aniseed-flavoured leaves and the seeds that follow the umbels of yellow flowers. Sow seed outdoors in spring and grow in full sun in fertile, well-drained soil. Water regularly to prevent plants from bolting (setting seed). Dill grows to 60–90 cm (2–3 ft) tall, but the dwarfer cultivars 'Bouquet', to 60 cm (24 in) tall, and 'Fernleaf', to 45 cm (18 in) tall, are pretty additions to summer containers. Use the leaves, fresh or dried, in cooking and harvest and dry the seeds for infusions. Dill is a trouble-free herb to cultivate.

Chive   Lime mint   Oregano   Hyssop

French tarragon One of the traditional *fines herbes* is French tarragon (*Artemisia dracunculus*), which is a hardy perennial with a rather upright habit, 1–1.2 m (3–4 ft) high, and narrow, aniseed-flavoured leaves. Small, yellow-green flowers appear in late summer but rarely set seed. Sow seed in spring or divide established plants in spring or autumn. Grow in full sun in well-drained, rich soil. Use the leaves to flavour egg and chicken dishes or add them to salads. Russian tarragon (*A. dracunculus* subsp. *dracunculoides*) is hardier and often sets seeds but has a less refined flavour than the species. Tarragon is susceptible to rust.

Borage The pretty, blue flowers of borage (*Borago officinalis*) are a popular addition to summer drinks and are often frozen into small ice cubes to add a splash of colour. The cucumber-flavoured leaves can also be added to drinks or used in salads. Borage is a hardy annual, easily grown from seed sown in spring. In the open garden it self-seeds freely, but in a container the flowerheads can be removed before they set seed. They are tolerant plants, growing to about 60 cm (24 in) high and 45 cm (18 in) across in any well-drained soil in full sun. Plants sometimes suffer from powdery mildew in hot, dry summers but are otherwise trouble free.

Camomile Roman camomile (*Chamaemelum nobile*) is a hardy, evergreen perennial, which grows to 15 cm (6 in) high and 45 cm (18 in) across. The white and yellow-centred, daisy-like flowers are borne in summer and it is these that are used to make camomile tea. Camomile can be grown from seed sown outdoors in spring, and established plants can be divided in spring. Grow in full sun in free-draining, fairly sandy soil. Infusions of the leaves are often used as a hair conditioner. The cultivar *C. nobile* 'Treneague', which does not bear flowers, is the plant to choose if you want a camomile lawn. All these plants are trouble free.

Coriander An aromatic hardy annual, coriander (*Coriandrum sativum*), which grows to 50–70 cm (20–28 in) high, has bright green leaves and in summer pale purple flowers. These are followed by yellowish seeds, which can be dried. Sow seed in spring and grow plants in rich, well-drained soil. If you want the leaves, position coriander in partial shade; if you want the seeds, make sure plants are in full sun. The fresh leaves are used to flavour soups and sauces and can be used as a garnish. The dried seeds are included in curries and pickles. Coriander is largely trouble free.

Cumin Cumin (*Cuminum cyminum*) is a half-hardy annual, growing only 30 cm (12 in) high. It has divided, dark green leaves and umbels of white to pale pink flowers in midsummer. The flowers are followed by greyish seeds. The rather bitter leaves are rarely eaten, but the seeds are an essential addition to curries and many Middle Eastern dishes. Sow seed under cover in early spring or outdoors after the last frosts. Grow in full sun in well-drained, rich soil. Cumin is trouble free.

Lemon grass A familiar ingredient in many Southeast Asian fish and meat dishes is lemon grass (*Cymbopogon citratus*), which is a tender perennial. It forms clumps of linear leaves and can get to 1.5 m (5 ft) high. In cold areas it can be kept in a conservatory or heated greenhouse in winter and moved outdoors in summer. Established plants can be divided in spring. As well as being a useful ingredient in cooking, the lemon-flavoured leaves can be used fresh to make tisanes. Plants are trouble free.

Fennel A hardy perennial, fennel (*Foeniculum vulgare*) is a tall herb, growing to 1.8 m (6 ft) or more. Sow seed under cover in early spring or outdoors in late spring, spacing plants to 45 cm (18 in). Fennel needs deep, fertile, well-drained soil and a position in full sun. The feathery, dark green leaves have a distinct aniseed flavour, and in summer small, yellow flowers are borne in flat umbels and are followed by small seeds. The cultivar 'Purpureum' has bronze-purple leaves. The leaves and seeds are often used to flavour fish dishes, and the leaves make pretty garnishes. Florence fennel

Fennel   Lavender   Parsley   Pineapple mint

(*F. vulgare* var. *dulce*) is a biennial, grown for the bulbous stalk base, which is treated as a vegetable or included raw in salads. Fennel is usually trouble free, although aphids and slugs can be a problem.

Hyssop A semi-evergreen, hardy, shrubby perennial, hyssop (*Hyssopus officinalis*) has aromatic leaves and spikes of purple-blue flowers in late summer. There are variants with white and pink flowers. Plants grow to 60 cm (24 in) high but tend to spread to about 1 m (3 ft). Sow seed in autumn and keep in a cold frame until spring, when plants should be moved to well-drained, fertile, alkaline soil and a position in full sun. The leaves have quite a strong flavour and should be used sparingly in meat dishes. Plants are trouble free.

Sweet bay A large evergreen shrub or tree, bay (*Laurus nobilis*) has aromatic, rather leathery leaves and, in spring, clusters of small, yellowish flowers, which are followed (on female plants) by black berries. In open ground bay trees can grow to 12 m (40 ft) tall, but container-grown plants can be pruned to keep them in proportion to the pot. They need deep, fertile, well-drained soil and some protection from cold, drying winds. They do best in full sun. The leaves are a familiar addition to marinades, soups and stews and are also a traditional part of bouquet garni. Scale insect can be troublesome, and bays are sometimes affected by powdery mildew and leaf spot.

Lavender There are dozens of cultivars of lavender (*Lavandula* spp.), with flowers in every possible shade of blue, pink and

purple as well as white. All have the wonderful, unmistakable fragrance and small, greyish leaves. Lavenders are evergreen shrubs, and *L. angustifolia* and its cultivars are hardy. Most other lavenders require some protection in winter. They need well-drained soil and a position in full sun. Some lavenders can grow large and quite leggy, so for containers look out for more compact cultivars, such as *L. angustifolia* 'Lavenite Petite', 'Princess Blue' or *L.* Bella Series in a range of colours. Use the fresh flowers in ice cream or vinegar or crystallize them to decorate cakes and puddings. Lavenders grown in well-drained soil in containers are usually trouble free.

Mint Mints (*Mentha* spp.) are among the easiest herbs to grow, and they are particularly well suited to growing in containers, which restrict their very invasive roots. They are hardy perennials, which can be grown from seed sown in spring. Established plants can be divided in spring or autumn. They do best in moisture-retentive but not too rich soil and prefer a position in full sun. Most have pinkish or purplish flowers, borne in spikes in summer above the aromatic leaves, which may be variegated. They range in habit from the creeping Corsican mint (*M. requienii*), which grows about 5 cm (2 in) high, to the vigorous eau-de-cologne mint (*M.* x *piperata* f. *citrata*), which can grow to about 50 cm (20 in) high and spread to 1 m (3 ft). Of all the available types, Moroccan mint (*Mentha spicata* var. *crispa* 'Moroccan') is perhaps the best culinary mint and one of the best for general all-round use. Rust and, in dry weather, powdery mildew may be problems.

| Purple basil | Golden curly marjoram | Sage | Lemon-variegated thyme |

Black cumin Also known as nutmeg flower and Roman coriander, black cumin (*Nigella sativa*) is a hardy annual, growing to 30 cm (12 in) high. It has finely divided leaves and small, white flowers in summer. These are followed by black seeds. Sow seed outdoors in autumn or spring in well-drained soil in full sun. The seeds can be used to flavour breads and pastries and are often included in curries and vegetable dishes. They can also be dried and used in infusions. These plants will self-seed and are trouble free.

Basil This most flavoursome of herbs is a tender annual or short-lived perennial. Basil (*Ocimum basilicum*), which grows to about 50 cm (20 in) tall, is cultivated for its aromatic leaves, which are included in many vegetable dishes, especially those containing tomatoes, pasta sauces and soups. Sow seed under cover in spring and plant out into fertile, well-drained soil after the last frosts. Grow in full sun for best results. Pick the leaves as plants come into flower. There are many cultivars, including *O. basilicum* var. *purpurascens* 'Purple Ruffles', which has purplish-black, crinkled leaves. Powdery mildew can be a problem in dry weather, and plants are sometimes infested with aphids.

Marjoram and oregano The perennials and subshrubs in the genus *Origanum* have aromatic leaves and spikes of, usually, pink flowers in summer. Sweet marjoram (*O. majorana*) is a half-hardy subshrub, to 60 cm (24 in) high, which is often treated as an annual in temperate areas. Oregano (*O. vulgare*) is a hardy perennial, growing to 45–90 cm (18–36 in) high; there

are many cultivars, with different leaf and flower colours. Sow seed of sweet marjoram under cover in spring and plant out after the last frosts. Sow seed of oregano in autumn or spring or divide established plants in spring. The fresh leaves of both marjoram and oregano are widely used in Italian and Greek cooking, and the dried leaves are suitable for infusions. Aphids are sometimes troublesome.

Parsley One of the most widely used of culinary herbs is parsley (*Petroselinum crispum*). Its bright green leaves are used to flavour soups, stews, butters and savoury dishes of all kinds and as a garnish. Flat-leaf parsley has a stronger flavour than the curly-leaved form. Parsley is a hardy biennial, to 60 cm (24 in) high. Sow seed outdoors from spring onwards for a succession of plants and water regularly until germination, which should be from 4–6 weeks after sowing. Transplant to 15 cm (6 in) apart into rich, moisture-retentive but well-drained soil. Grow in full sun or partial shade. Carrot fly larvae and celery fly larvae can damage roots and leaves, respectively. Hamburg parsley (*P. crispum* var. *tuberosum*) is grown for its parsnip-like roots, which are harvested in autumn and early winter. Sow seed outdoors in alkaline, fertile soil in partial shade; plants mature in about 30 weeks and have a better flavour if left in the ground rather than being lifted and stored. Parsnip canker is the main problem.

Rosemary Rosemary (*Rosmarinus officinalis*) is an evergreen shrub, with aromatic leaves and, in summer, blue, pink or white flowers. Although it is not absolutely hardy, the species and most of the numerous cultivars will withstand frosts for short periods. Although rosemary can be grown from spring-sown seed, it is easier to buy plants and transfer them to well-drained, moderately fertile soil and a position in full sun. The species can grow to 1.5 m (5 ft) tall, and the other most widely available cultivar, 'Miss Jessopp's Upright', is also tall growing. Better for containers are the smaller 'Severn Sea', to 1 m (3 ft) tall, or even the low-growing (but less hardy), 'Prostratus', which gets to 15 cm (6 in) high. The leaves are a traditional accompaniment for lamb but can also be added to cakes and preserves or used to infuse olive oil. Plants are usually trouble free.

Sage This large genus, *Salvia*, includes the ornamental bedding annuals as well as the perennial and shrubby culinary herbs. Common sage (*S. officinalis*) is a hardy, shrubby, evergreen perennial, to 80 cm (32 in) high, with velvety, grey-green leaves and purple-blue, pink or white flowers. Numerous cultivars have been developed, including 'Tricolor', which has grey-green leaves with cream and pink margins, and the compact 'Kew Gold', which has green-flecked, golden-yellow leaves. The species (*S. officinalis*) can be grown from spring-sown seed but the cultivars should be grown from cuttings taken in summer or purchased as plants. Grow them in well-drained soil in full sun. The leaves can be used to make tea or to flavour meat dishes and cheeses. These sages can be affected by leafhoppers.

Summer savory A hardy annual, summer savory (*Satureja hortensis*), which grows to 25–38 cm (10–15 in) tall, has narrow green leaves and whorls of lilac to white or purple flowers in summer. Sow seed under cover in late winter or outdoors in spring and grow plants in well-drained, fertile, alkaline soil in full sun. The leaves, an ingredient in the traditional *herbes de Provence*, can be used fresh in meat dishes and to flavour stuffings. Winter savory (*S. montana*) is a hardy dwarf subshrub, with pink-purple flowers. Use the leaves of winter savory in the same way as those of summer savory. Both plants are trouble free.

Thyme The large genus, *Thymus*, includes several small, evergreen perennials and subshrubs that often do better in the free-draining conditions that can be provided in a container than in the open garden. Garden or common thyme (*T. vulgaris*), a hardy subshrub, grows to 30 cm (12 in) tall and has aromatic, grey-green leaves and purple to white flowers in late spring and early summer; the cultivar 'Silver Posie' has white-edged leaves. *T. citriodorus* 'Golden King', to 25 cm (10 in) tall has yellow-edged leaves. Thymes are readily raised from summer cuttings, but it is usually easier to buy plants of the cultivar of your choice. Plant in well-drained, neutral to alkaline soil in full sun. The small leaves of thyme are included in bouquet garni and as an ingredient in many classic French dishes. The leaves are also used in stuffings, casseroles and marinades. The plants are trouble free.

Fenugreek The half-hardy annual fenugreek (*Trigonella foenum-graecum*) can be easily grown from seed sown outdoors in spring. Plants, which grow 50–60 cm (20–24 in) high and to 45 cm (18 in) across, need fairly deep, rich, free-draining soil and a position in full sun. The aromatic leaves, which can be used as a vegetable or dried for infusions, have three, toothed leaflets. The yellow-white flowers are borne in late spring to summer and are followed by pods, which contain yellowish seeds. The ground seeds are an ingredient in curries and pickles. These are trouble-free plants. If you have some seed left over and some spare ground in the vegetable plot, use fenugreek as a nitrogen-fixing green manure, sowing in late spring for digging in in late summer.

Edible flowers

Agastache There are numerous species and cultivars of agastache. The most commonly used culinary variety is anise hyssop (*A. foeniculum*), which will grow 1–1.5 m (3–5 ft) tall and bear spikes of small, purplish-blue flowers. All agastaches require well-drained soil and a sunny position, flowering from mid-summer to early autumn. Species can be raised from seed sown under cover in spring, whilst cultivars should be propagated by division in spring. These are sturdy, largely trouble-free plants, which sometimes suffer from powdery mildew in dry weather.

Pot marigold Fast-growing, hardy annuals, pot or English marigolds (*Calendula officinalis*) have aromatic leaves and single or double, daisy-like flowers in shades of yellow, orange, cream and gold in summer and autumn. There are numerous cultivars of this, the names often indicating the flower colour: 'Orange King', 'Lemon Queen', 'Pink Surprise' and 'Greenheart Orange', for example. Most are 45–60 cm (18–24 in) high. They grow easily from spring- or autumn-sown seed and will tolerate fairly poor soil as long as it is well drained. Deadheading regularly will encourage new blooms. The petals can be used (instead of saffron) to colour rice, and they are also tasty in butters, cheeses and cakes or sprinkled over salads. Plants are susceptible to aphids and powdery mildew. Most grow 45–60 cm (18–24 in) high, but there are also several dwarfer, more compact forms.

Cornflower A hardy annual, cornflower (*Centaurea cyanus*) has dark blue flowers from late spring to mid-summer. Plants grow to 80 cm (32 in) tall and have grey-green leaves. Smaller cultivars, better suited to containers, include the Baby Series, which grow to 30 cm (12 in) tall and have blue, pink or white flowers. Sow seed outdoors in autumn or spring and plant seedlings into well-drained soil in full sun. The flowers can used fresh in salads or dried and included in pot-pourri. Although largely trouble free, cornflowers are susceptible to powdery mildew in hot, dry weather.

Clove pink Clove pink or gillyflower (*Dianthus caryophyllus*) is a hardy, evergreen perennial, to 50 cm (20 in) tall, with grey-green leaves and pink to purple, clove-scented flowers in summer. Sow seed under cover in spring or buy plants of named cultivars and plant them in fertile, well-drained soil in full sun. Deadhead regularly to encourage new flowers. The fresh flowers can be added to salads, although they are mostly used as flavouring for syrups and liqueurs. Powdery mildew and root rot can be problems.

Sunflower Most sunflowers (*Helianthus annuus*) are too tall for containers, some growing to 3 m (10 ft) or more, but some shorter cultivars of these hardy annuals have been developed, including the compact 'Teddy Bear', to 90 cm (3 ft) high, and 'Big Smile', to 40 cm (16 in) high. Sow seed outdoors in spring and grow in fertile, moisture-retentive but well-drained soil in full sun. Whilst they are still green, the flowerbuds are edible and can be fried in butter, whilst the seeds can be eaten raw or roasted. In hot, dry summers powdery mildew can be a problem. Plants should be protected from slugs and snails.

Daylily Hardy herbaceous perennials, daylilies (*Hemerocallis* spp. and cultivars) form clumps of narrow strap-shaped leaves and strong stems bearing a profusion of lily-like blooms in summer. Colours range from deep orange-red, through yellow and pink to creamy white and each flower lasts just a day. Daylilies thrive in moisture-retentive but free-draining soil in full sun or partial shade and are best divided in spring every three years. The buds and the petals are edible, having a sweet flavour reminiscent of water chestnuts and a crunchy lettuce-like texture. Like many edible flowers they should be eaten in moderation. Use in salads and stir-fries, and for garnishing. There are thousands of named varieties, ranging in height from 30 cm (12 in) to 1.2 m (4 ft). Daylilies are generally trouble free, although slugs and snails are atttracted to the tender new shoots in spring.

Viola   Nasturtium   Pot marigold   Cornflower

Scented-leaved pelargonium This group of tender, shrubby, evergreen perennials have highly aromatic foliage, which releases its strong fragrance when the plants are brushed. They are quite different from other members of the genus *Pelargonium*, such as the ivy-leaved and zonal forms, which are popular bedding subjects. Scented-leaved pelargoniums have small, single flowers, which may be pink, purple or white, and mid-green leaves, which may be variegated with gold or silver. Buy plants from a good nursery or garden centre and plant them out in well-drained, fertile, neutral to alkaline soil. Deadhead to encourage new flowers, and lift before the first frosts to overwinter in a dry, frost-free greenhouse. The leaves of *P. citronellum*, lemon pelargonium (*P. crispum*) and apple pelargonium (*P. odoratissimum*) can be used to make tea, but these plants are mostly grown for their fragrant oils. Plants are susceptible to grey mould and aphids.

Primrose The most welcome of spring flowers are primroses (*Primula vulgaris*), which bear pale yellow, fragrant flowers in spring above the rosettes of bright green leaves. They are hardy perennials, growing to 20 cm (8 in) high. Sow seed in late winter or early spring and grow in partial shade in rich, well-drained but moisture-retentive soil. Divide established plants regularly to maintain their vigour. The flowers and leaves can be added to salads, and the flowers are pretty garnishes for delicate puddings. Plants are susceptible to grey mould and slugs.

Nasturtium The vividly coloured flowers of nasturtiums (*Tropaeolum majus*) are a frequent addition to salads, but the peppery leaves can also be eaten in salads, and the seeds can be pickled and used as a substitute for capers. There are many hybrids, which can be easily grown from seed sown outdoors in spring or started under glass for earlier flowers. These annuals are naturally climbing or trailing plants, but there are an increasing number of dwarf or compact cultivars available. Nasturtiums do best in full sun and moisture-retentive but free-draining soil; if it is too rich, however, plants will produce leaves at the expense of flowers. Look out for compact varieties, such as 'Empress of India', which gets to 30 cm (12 in) tall and has scarlet flowers, or plants in the Tom Thumb Series, which have bright yellow, orange, red and salmon-pink blooms. Blackfly and cabbage white butterfly caterpillars can devastate nasturtiums. Be vigilant and remove the pests by hand.

Viola and pansy The annuals and perennials in the genus (*Viola*) have been widely hybridized to produce numerous cultivars. If you require only a few plants it is most sensible to buy plantlets, although seed can be sown in late winter for early spring and summer flowers or in summer for winter flowers. Grow them in moisture-retentive but well-drained, fertile soil in sun or partial shade. The dainty sweet violet (*V. odorata*) bears fragrant, dark purple or white flowers from late winter to early spring. These flowers are often used as a garnish for desserts or fresh in salads, and they can be crystallized or frosted (with egg white and sugar) to decorate cakes. Slugs and snails can be a problem, as can powdery mildew in dry weather.

Index

A

acid composts 15
Agastache
    growing instructions 156
    projects 78
Ajuga reptans 120
amaranth 141
American cress 68
annual flowers 12
aphids 12, 27
apples
    growing instructions 21, 148
    projects 112
    recipe 112
asparagus peas
    growing instructions 144
    projects 86
aubergines
    growing instructions 146
    projects 77, 88
    recipes 77, 88

B

backgrounds, siting containers
    11
banana plants
    growing instructions 25, 148
    projects 124
basil
    growing instructions 154
    projects 98
bay, sweet
    growing instructions 21, 25,
      153
    projects 102
beans, sowing 19
    see also broad beans, runner
      beans etc
beetroot
    growing instructions 18,
      142
    projects 37, 64
    recipe 64
Begonia 112
biological pest control 27, 29
black-eyed Susan 38
blackcurrants 18, 150
blackfly 27
blight 28
blueberries
    growing instructions 25,
      148–9
    projects 114
    recipe 114

borage
    growing instructions 152
    projects 50
borecole 139
botrytis 28
brassicas 19, 27
    see also cabbages

C

cabbages
    growing instructions 25, 138
    projects 100
    recipe 100
Calendula officinalis
    growing instructions 156
    projects 67, 68
Calibrachoa 94
camomile
    growing instructions 152
    projects 126
    recipe 126
Cape gooseberries
    growing instructions 149
    projects 134
Carex comans 78
carrots
    growing instructions 18, 142
    projects 37
    recipes 37, 91, 124
caterpillars 27
cauliflower
    growing instructions 138
    projects 73
    recipe 73
cell trays, sowing seeds 19
Centaurea cyanus
    growing instructions 12, 156
    projects 67
ceramic pots 13
chard
    growing instructions 138
    projects 73, 81, 92
    recipe 92
chicken, recipes 52, 94, 104
chemical pest control 29
cherries 21
chilli peppers
    growing instructions 147
    projects 82, 104
    recipes 82, 104
Chinese cabbage
    growing instructions 139
    projects 108
Chinese chives 60, 107
chives 151
choosing containers 13–14
citrus fruits 26, 149
clary, annual 88

climbing plants 12, 21
clove pinks 156
coir, soil-less composts 15
companion planting 12
composts
    additives 15
    choosing 15
    mulching 25,
    planting containers 16
    sowing seeds 19
containers
    choosing 13–14
    grouping 11
    siting 11
concrete containers 14
coriander
    growing instructions 152
    projects 60, 78, 82, 107
corn salad 141
cornflowers
    growing instructions 12,156
    projects 67
courgettes
    growing instructions 146
    projects 58, 124
    recipe 124
cranberries
    growing instructions 25, 149
    projects 114
cucumbers 18, 19, 146–7
cumin 152, 154
cut-and-come-again crops 34
cuttings 18

D

daylilies 156
deadheading 21
Dianthus
    growing instructions 156
    projects 122
dill
    growing instructions 151
    projects 52, 60
diseases 20, 28–9
    see also blight, rust etc
division 18
dolichos beans 145
drainage 16
drip systems 23
dwarf beans
    growing instructions 21,
      144–5
    projects 46, 63, 77, 86,

E

edible flowers 67, 156–7
    see also cornflowers,
nasturtiums etc

eggplant *see* aubergines
endive 138–9
ericaceous composts 15

F

F1 hybrids 18
feeding plants 26
fennel
    growing instructions 18,
      152–3
    projects 97
    recipe 97
fennel, bronze 97
fenugreek 155
fertilizers 17, 26,
fibreglass containers 14
fish, recipes 78, 102
flowers
    deadheading 21
    edible 156–7
French beans
    growing instructions 144–5
    projects 63, 86
    recipe 86
fruit 12, 21, 148–50
    see also apples, pears etc

G

gages 21
garlic 143
garlic chives
    growing instructions 151
    projects 60, 107
germination, seeds 18, 19
gherkins 146–7
glazed ceramic pots 13
gooseberries 18
gourds
    growing instructions 12, 21,
      147
    projects 130
grape vines
    growing instructions 149
    projects 120
    recipe 120
greenfly 27
grey mould 28
grit 15, 25
grouping containers 11

H

hanging baskets, planting 17
Helianthus annuus
    growing instructions 156
    projects 67
Hemerocallis 156
herbs 18, 21, 151–5
    see also basil, thyme etc

hops 130
hyacinth beans 145
hybrids, F1 18
hygiene 16, 20, 29
hyssop
  growing instructions 153
  projects 78, 84

I
improvised containers 14
insecticides 29
insects 12
Ipomoea
  growing instructions 12
  projects 70
irrigation systems 23

K
kale
  growing instructions 139
  projects 60, 73, 100
kohl rabi
  growing instructions 143–4
  projects 91
  recipe 91

L
land cress
  growing instructions 141
  projects 68
lavender
  growing instructions 18, 25,
    153
  projects 67
leaf beet
  growing instructions 138
  projects 60, 81, 92
leaf miners 27
leafhoppers 27
leeks
  growing instructions 144
  projects 64
lemon balm 18
lemon grass
  growing instructions 152
  projects 82
lemon verbena
  growing instructions 151
  projects 126
lemons
  growing instructions 149
  projects 117
  recipes 50, 117, 120
lettuces
  growing instructions 19, 25,
    139
  growing with flowers 43, 49,
    64

projects 37, 43, 49
  recipes 43, 49
lime-hating plants 15
liners, hanging baskets 17
lobelia 49
Lotus maculates x *berthelottii* 104

M
marigolds
  growing instructions 12, 29,
    156
  French marigolds 12, 29
  pot marigolds 67, 68, 156
  projects 67, 68
marjoram
  growing instructions 154
  projects 74, 97
melons 18
metal containers 13, 14
mildew, powdery 28
mint
  growing instructions 10, 18,
    153
  projects 81, 84
  recipes 81, 84, 91
mizuna greens
  growing instructions 139–40
  projects 44
morning glory
  growing instructions 12
  projects 70
mulching 25
mustard
  growing instructions 141
  projects 44

N
nasturtiums
  growing instructions 157
  projects 60, 67, 94
nectarines 21
nitrogen 26

O
olive trees
  growing instructions 25
  projects 98
onions
  growing instructions 25, 144
  projects 54
  recipe 78
orache 141
oranges 149
oregano
  growing instructions 154
  projects 74, 97
organic gardening 29
Oriental greens 139–40

ornamental plants, mixing with
  edible plants 12
overwatering 23

P
pak choi
  growing instructions 140
  projects 44
  recipe 44
pansies 157
para cress
  growing instructions 141
  projects 97
parrot's beak 104
parsley
  growing instructions 18, 154
  projects 38, 52, 60, 68, 84,
    107
passionflowers
  growing instructions 21,
    149–50
  projects 134
passionfruit 149–50
  recipes 112, 134
pasta, recipe 58
peaches 21
pears
  growing instructions 21, 150
  projects 133
  recipes 54, 108, 133
peas
  growing instructions 19, 145
  projects 33
peat 15
Pelargonium
  growing instructions 102, 157
  projects 117
pennyroyal 81
peppermint
  growing instructions 153
  projects 84, 126
peppers
  growing instructions 18, 19,
    147
  projects 82, 94
  recipes 64, 94
perennial flowers 12
perlite 15
pests 12, 20, 27–8, 29
  see also aphids, slugs etc
pinks 156
pizza, recipes 34, 74
planting
  hanging basket 17
  pot 16
plastic pots 13
Plecostachys serpyllifolia 64
plums 21

poisonous plants 12
pollination 12
popcorn 129
potash 26
potatoes
  growing instructions 142
  projects 84
  recipe 84
powdery mildew 28
primroses 157
propagation 18–19
pruning 20–1
pumpkin 147
purslane 141

R
radicchio
  growing instructions 140
  projects 92, 108, 133
  recipe 108
radishes 142–3
red cabbage
  projects 100
  recipe 37
red spider mite 27
redcurrants
  growing instructions 150
  projects 122
rhubarb chard 139
risotto, recipe 70
rocket
  growing instructions 140
  projects 74, 98, 107
  recipe 64
roof gardens 13
rosemary
  growing instructions 18,
    154–5
  projects 102
runner beans
  growing instructions 12, 21,
    145
  projects 46, 70, 77
  recipe 46
rust 28–9

S
sage 88, 155
salad, recipes 37, 43, 52, 54, 64,
  67, 86, 91, 97, 98, 107
salad burnet
  growing instructions 140
  projects 50
salad leaves
  growing instructions 18, 26,
    140–1
  projects 34
Salvia elegans 117

Salvia viridis 88
savory
 growing instructions 155
 projects 78, 82
seafood, recipe 44
seeds, sowing 18–19
self-watering containers 24
shallots 25, 144
siting containers 11
slugs 25, 27, 29
snails 25, 27
soil-based composts 15
soil-less composts 15
sorrel
 growing instructions 141
 projects 68
sowing seeds 18–19
spinach
 growing instructions 141
 projects 54
 recipe 54
spinach mustard 140
squashes
 growing instructions 18, 147

projects 130
 recipe 130
sterilizing equipment 20
Stipa tenuissima 78
stir-fry, recipes 33, 46
stone containers 14
stone fruits 21
stoneware pots 13
strawberries
 growing instructions 18, 150
 projects 50, 60, 118, 129
 recipes 50, 129
sunflowers
 growing instructions 156
 projects 46, 67
supports 21
Sutera cordata 64
sweet peas 33
sweetcorn
 growing instructions 147
 projects 63, 129
 recipe 63

T
tarragon, French
 growing instructions 152
 projects 52
 recipe 52
terracotta pots 13
terrazzo containers 14
thyme
 growing instructions 18, 155
 projects 40, 52, 60, 78, 84, 91
 recipe 40
timber containers 14
tomatoes
 fertilizers 26
 growing instructions 147
 in hanging baskets 38
 pests 12
 pinching out sideshoots 20
 projects 58, 74, 133
 recipes 38, 74, 77, 81, 82, 86
 sowing 18, 19
 supports 21
topiary, trimming 21
training plants 21

tulips 43
turnips 143

V
vegetables 138–47
 see also cabbages, leeks etc
Verbena 64
vermiculite 15
vine weevils 27, 29
violas
 growing instructions 157
 projects 43, 60, 67

W
water-retaining granules 15, 23
watering 19, 22–4
waterlogging 23
wetting agents 23
white currants 150
whitefly 12, 28, 29
winter cress 141
wooden containers 14
woodlice 28

Acknowledgements

Executive Editor Sarah Ford

Managing Editor Clare Churly

Editor Lydia Darbyshire

Executive Art Editor Karen Sawyer

Designer Miranda Harvey

Illustrator Box 68

Photographer Freia Turland

Senior Production Controller Martin Croshaw

Author's acknowledgements
With endless thanks to my mum, Carol, for her unfailing love and support, to Judy Holbrook and Andy Luft for their invaluable help, advice and friendship, and to all at Hamlyn, especially Sarah Ford and Clare Churly, for their patience, expertise and guidance. My gratitude also goes to those individuals and companies who have helped along the way. Lastly, but certainly not least, thank you Freia for remaining calm and for producing such gorgeous images.

With grateful thanks to the following companies without the support of whom this book could not have been produced.

Seed companies
Chiltern Seeds, Bortree Stile, Ulverston, Cumbria, LA12 7PB
Tel: 01229 581137
www.chilternseeds.co.uk

Samuel Dobie & Son, Long Road, Paignton, Devon, TQ4 7SX
Tel: 0870 112 3625
www.dobies.co.uk

Mr Fothergill's Seeds, Kentford, Suffolk, CB8 7QB
Tel: 0845 166 2511
www.mr-fothergills.co.uk

Plants of Distinction, Abacus House, Station Yard, Needham Market, Suffolk, IP6 8AS
Tel: 01449 721720
www.plantsofdistinction.co.uk

Suttons Seeds, Woodview Road, Paignton, Devon, TQ4 7NG
Tel: 0870 220 0606
www.suttons-seeds.co.uk

Thompson and Morgan (UK) Ltd, Poplar Lane, Ipswich, IP8 3BU
Tel: 01473 695225
www.thompson-morgan.com

Container suppliers
C H Brannam Ltd, Roundswell Industrial Estate, Barnstaple, Devon, EX31 3NJ
Tel: 01271 343035

Iota Garden and Home Ltd, Wick Road, Wick St Lawrence, North Somerset, BS22 7YQ
Tel: 01934 522617
www.iotagarden.com

Woodlodge Products Ltd, Woodlodge, Holloway Hill, Chertsey, Surrey, KT16 0AE.
www.woodlodge.co.uk.

Other suppliers
Cadbury Garden and Leisure, Smallway, Congresbury, Bristol, BS49 5AA.
Tel: 01934 875700.
www.g-l.co.uk.

The National Collection of Passiflora, Greenholm Nurseries Ltd, Lampley Road, Kingston Seymour, Clevedon, North Somerset, BS21 6XS.
Tel: 01934 833350.